MARINEFF International Conference: From Materials and Infrastructures to Marine Ecosystem - Interactions and New Approaches

IOP Conference Series: Materials Science and Engineering
Volume 1245

Epron, France
3 - 5 May 2022

Editor:

Mohamed Boutouil

ISBN: 978-1-7138-6012-9
ISSN: 1757-8981

Printed from e-media with permission by:

Curran Associates, Inc.
57 Morehouse Lane
Red Hook, NY 12571

Some format issues inherent in the e-media version may also appear in this print version.

Institute of Physics
Dirac House, Temple Back
Bristol BS1 6BE UK

Phone: 44 1 17 929 7481
Fax: 44 1 17 920 0979

techtracking@iop.org

Additional copies of this publication are available from:

Curran Associates, Inc.
57 Morehouse Lane
Red Hook, NY 12571 USA
Phone: 845-758-0400
Fax: 845-758-2633
Email: curran@proceedings.com
Web: www.proceedings.com

TABLE OF CONTENTS

IOP Conf. Series: Materials Science and Engineering 1245 (2022) 011001 doi:10.1088/1757-899X/1245/1/011001

FROM MATERIALS AND INFRASTRUCTURES TO MARINE ECOSYSTEM: INTERACTIONS AND NEW APPROACHES

MARINEFF
MARine INfrastructures EFFects

May 3rd-5th, 2022

ESITC Caen - France

Edited by
Dr. Mohamed BOUTOUIL,
Research Director
ESITC Caen

With the support of the MARINEFF partnership:

With the support of Interreg Va France (Channel) England program:

And the sponsorship of AFGC:

IOP Conf. Series: Materials Science and Engineering 1245 (2022) 011001 doi:10.1088/1757-899X/1245/1/011001

- PREFACE -

As a lead partner of the European Interreg Project MARINEFF, ESITC Caen organized 'MARINEFF Conference: From materials and infrastructures to marine ecosystem: interactions and new approaches' in Caen (Normandy, France) from the 3rd to 5th May 2022.

The MARINEFF project was selected under the European cross-Broder Cooperation Programme INTERREG V/A between France (Channel) England, co-funded by the ERDF. The project brings together 9 French and English academia and industrial partners. The main MARINEFF project goal is to enhance and protect coastal and transitional water ecosystems in cross-border Channel regions by developing and implementing new biomimetic concrete infrastructures that improves the initial water ecological status.

The congress papers and contributions focused on the following main topics: 1) Marine infrastructures (incl.artificial reefs) eco-engineering and nature-based solutions: ecological design, engineering and technology, 2) Impacts of marine infrastructures on biodiversity and coastal functioning – Monitoring strategies and protocols, 3) Interactions (physical, mechanical, chemical and biological) between materials (concrete, steel, wood, composite...) and the marine environment and biodiversity 4) Stakeholders engagement and collaborative approaches with respect to coastal environment and economy and 5) MARINEFF main results.

Within these topics, more than thirty communications were selected. During the two days conference, the research and technical results have been presented in oral communications and two valuable invited lectures have been given, respectively by Dr Joe Ironside, (Aberystwyth University, UK) and Dr Sylvain Pioch (Université PV Montpellier, France), on 'Ecostructure project: An ecologically sensitive approach to coastal and marine infrastructure for the Irish Sea.' and 'Overview of marine infrastructures eco-engineering and natured based solutions : from concept to main results applications' respectively.

The communications edited in this proceeding consist of peer-reviewed papers. We are grateful to the Scientific Committee members for their valuable advice and comments. Thanks to the speakers, we hope, and we are quite sure, that the conference gave new impetus to the research work of the participants and allowed the researchers and professionals to discuss and to investigate the various topics and challenges related to biomimetic concrete infrastructures.

ESITC Caen and MARINEFF partners are grateful to the European Regional Development Fund (INTERREG V/A France (Channel) England) for co-funding and supporting this conference.

Warm and special thanks and complements to the Organizing Committee and the ESITC Caen staff for their helpful efforts in organizing the MARINEFF Conference.

Caen, 3rd May, 2022

Dr. M. BOUTOUIL, ESITC Caen.

MARINEFF Project Coordinator

Deputy Director and Research Director, ESITC Caen

IOP Conf. Series: Materials Science and Engineering 1245 (2022) 011001 doi:10.1088/1757-899X/1245/1/011001

- PREFACE -

L'ESITC Caen, en tant que chef de file du projet Européen MARINEFF, a organisé à Caen (France), du 3 au 5 mai 2022, le "Congrès MARINEFF, des matériaux aux infrastructures pour l'écosystème marin : interactions and nouvelles approches".

Le projet MARINEFF a été sélectionné dans le cadre du Programme européen de coopération transfrontalière INTERREG V/A France (Manche) Angleterre, cofinancé par le FEDER et rassemble 9 partenaires français et britanniques (académiques et industriels). Les objectifs du projet MARINEFF sont multiples : l'amélioration de la biodiversité et de la production de l'écosystème marin de la Manche, et le développement et l'évaluation d'infrastructures maritimes biomimétiques en béton.

Les communications du congrès se sont concentrées sur les principaux thèmes suivants : 1) Conception, ingénierie et technologie écologiques des infrastructures marines (y compris récifs artificiels), 2) Impacts des infrastructures marines sur la biodiversité et le fonctionnement des côtes - Stratégies et protocoles de suivi, 3) Interactions (physiques, mécaniques, chimiques et biologiques) entre les matériaux (béton, acier, bois, composite...) et le milieu marin et la biodiversité, 4) Engagement des parties prenantes et approches collaboratives en ce qui concerne l'environnement côtier et l'économie et 5) Principaux résultats du projet MARINEFF.

Dans le cadre de ces sujets, plus d'une trentaine de communications ont été sélectionnée. Pendant les deux jours de congrès, les récents résultats techniques et scientifiques ont été présentés lors de communications orales et deux conférences invitées ont été données par Dr Joe Ironside, (Aberystwyth University, UK) et Dr Sylvain Pioch (Université PV Montpellier, France), respectivement sur les sujets suivants : 'Ecostructure project: An ecologically sensitive approach to coastal and marine infrastructure for the Irish Sea.' and 'Overview of marine infrastructures eco-engineering and natured based solutions : from concept to main results applications'.

Les articles édités dans ces actes ont été examinés par des rapporteurs pour être présentés lors des sessions orales. Nous tenons à remercier vivement les membres du Comité Scientifique pour leurs conseils et commentaires. Nous espérons que le congrès, grâce aux communications orales et aux articles édités a su et pu donner une nouvelle impulsion aux travaux de recherche des participants et a permis aux chercheurs et aux professionnels d'échanger sur les différents aspects et défis des infrastructures maritimes biomimétique en béton.

L'ESITC Caen et les partenaires du projet MARINEFF remercient les Fonds Européen de Développement Régional (INTERREG V/A France (Manche)/Angleterre) pour le cofinancement de l'organisation du congrès.

Nos chaleureux remerciements et compliments vont également au Comité d'Organisation et au personnel de l'ESITC Caen pour sa disponibilité et ses efforts dans l'organisation du Congrès MARINEFF.

Caen, le 3 mai 2022

Dr. M. BOUTOUIL, ESITC Caen.

Coordinateur Projet MARINEFF

Directeur Délégué et Directeur Scientifique, ESITC Caen

IOP Conf. Series: Materials Science and Engineering 1245 (2022) 011001 doi:10.1088/1757-899X/1245/1/011001

- MARINEFF INTERNATIONAL CONFERENCE -

ORGANIZING COMMITTEE

Chair:

BOUTOUIL Mohamed

Members:

BOURGUIBA	Amal
BOUTOUIL	Mohamed
COUBE	Marie-Caroline
DUFEU	Matthieu
GERAULT	Aurélie
LEBRUN	Jérôme
LEPAGE	Mathieu
SEBAIBI	Nassim

SCIENTIFIC COMMITTEE

Chair:

BOUTOUIL Mohamed

Members:

ANDERSEN	Jens	GUEGUIN	Marielle
AUZAS	Mickaël	GUIRAUD	Patrick
BERTRON	Alexandra	HERBERT	Roger
BOURGUIBA	Amal	HUBERT	Philippe
BOUTOUIL	Mohamed	IWANAMI	Mitsuyasu
CLAQUIN	Pascal	JENSEN	Antony
COLLINS	Ken	SALGUES	Marie
DAUVIN	Jean-Claude	SEBAIBI	Nassim
DUBOIS-BRUGGER	Isabelle	WILLEMY	Charles
FABIEN	Aurélie	YSNEL	Frédéric
FEUNTEUN	Eric		

- PARTNERS OF THE MARINEFF PROJECT -

ESITC CAEN (LEADER)

Contact: Dr. Mohamed BOUTOUIL
1 rue Pierre et Marie Curie
14610 EPRON
France
www.esitc-caen.fr

VINCI CONSTRUCTION MARITIME ET FLUVIAL (VCMF)

Contact: Mickaël AUZAS
1 rue Pierre et Marie Curie
14610 EPRON
France

NATIONAL MUSEUM OF NATURAL HISTORY

Contact: Dr. Eric FEUNTEUN
Marine Station of Dinard
38 rue du Port Blanc
35800 DINARD
France
www.mnhn.fr/le-museum/

UNIVERSITY OF SOUTHAMPTON

Contact: Dr. Ken J. COLLINS
National Oceanographic Centre
Southampton, SO14 3ZH
United Kingdom
www.southampton.ac.uk/oes

PORTS DE NORMANDIE

Contact: Bertrand MARSSET
3 rue René Cassin
14280 SAINT-CONTEST
France
www.pna-ports.fr

BOURNEMOUTH UNIVERSITY

Contact: Pr. Roger HERBERT
Fern Barrow
Poole BH12 5BB
United Kingdom
www.bournemouth.ac.uk

TPC (VINCI)

Contact: Arnaud PASSELAC
Victorien Mouchel
160 rue Sauxmarais
BP 7
50110 TOURLAVILLE
France
www.entreprise-tpc.fr

UNIVERSITY OF EXETER

Contact: Dr. Jens ANDERSEN
Cambourne School of Mines
Penryn Campus
Treliever Road
Penryn, Cornwall, TR10 9EZ
United Kingdom
www.exeter.ac.uk

UNIVERSITE DE CAEN

Contact: Pr. Pascal CLAQUIN
Esplanade de la paix
CS 14032
14032 CAEN cedex
France
www.unicaen.fr

Peer Review Statement

All papers published in this volume have been reviewed through processes administered by the Editors. Reviews were conducted by expert referees to the professional and scientific standards expected of a proceedings journal published by IOP Publishing.

- **Type of peer review:** Undefined
- **Conference submission management system:** Morressier
- **Number of submissions received:** 34
- **Number of submissions sent for review:** 33
- **Number of submissions accepted:** 29
- **Acceptance Rate (Submissions Accepted / Submissions Received × 100):** 85.3
- **Average number of reviews per paper:** 1.32
- **Total number of reviewers involved:** 15
- **Contact person for queries:**
  Name: GERAULT
  Email: aurelje.gerault@esitc-caen.fr
  Affiliation: Ecole Supérieure d'Ingénieurs des Travaux de la Construction - Research laboratory

Can port concrete infrastructures be optimized to promote algal and macrofaunal colonisation in the marine intertidal zone? Case study of Port Haliguen (Brittany, France)

M Lapinski[1*], M Perrot[1] and P Sauleau[2]

[1] Seaboost, 889 Rue de la Vieille Poste, 34000 Montpellier, France
[2] Laboratoire de Biotechnologie et Chimie Marines, Univ. Bretagne-Sud, EA 3884, LBCM, IUEM, F-56100, Lorient, France

*Corresponding author email: Matthieu.LAPINSKI@seaboost.fr

Abstract. Various studies highlight the efficiency of working on the nature of port infrastructures (materials, structure design for water retention and creation of microhabitats) in order to improve ecological functions in and around port areas. The aim of this work which was carried out at Port Haliguen (Brittany, France) was to study the colonisation of different types of concrete formulations, including 3D-printed concrete, assumed as more favourable for sessile marine life due to their intrinsic characteristics (surface pH, roughness, porosity). During the first year, no differences were shown between the settlement of algae and macrozoobenthos on each of the formulations regardless of the water level. Percentages of coverage were extremely high after only 5 months and mean diversities were not different. However, the finer grain porous concrete showed a great potential to promote the settlement of the engineer species *Fucus vesiculosus* on vertical seawalls. The 3D-printed concrete formulation appears to date to be compatible with marine life colonisation. The advantages of this new technologies to create complex habitats could be of great interest applied to the integration of large scale eco-engineered strategies in urbanized shoreline areas such as ports in Brittany.

1. Introduction

The 'ocean sprawl' of the coast coupled with population increase represents an unprecedented threat to the diversity of marine species and their ecological connectivity [1, 2]. These threats of human origin intervene at difference scales in the living world and result in the alteration of the composition of marine microbial communities [3], the destruction of natural habitats and their replacement by artificial structures [4, 5], as well as noise pollution [6] and light pollution [7], that is still underestimated. This is all the more cause for concern in that the biodiversity and the richness of marine habitats is concentrated along the coasts [8].

The artificialisation of the marine environment involves in particular the implantation along the coastline of port infrastructures such as seawalls, dykes, piles or breakwaters for protection against the rise in sea level, wind farms, artificial reefs and insubmersible tanks for oyster production. These man-made structures, generally of concrete, are not designed to recreate biodiversity which would in many cases be different from that which occurs in the nearby natural habitats [9, 10, 11]. Nevertheless, in the face of the erosion of marine biodiversity, it would appear that these man-made structures might henceforth be considered as a real ecological asset by playing a role in the restoration of the marine environment [12], in particular via the addition of engineer species and by the increase in their abundance.

In the port environment, most of the dykes, seawalls, piles, etc. are lacking in physical complexity. This lack of surface complexity conditions and limits the composition of the microbial biofilm [13] and in consequences the recruitment, settlement and dispersal of marine invertebrates as well as the trophic interactions between species [14, 15, 16]. This physical homogeneity ultimately results in a biological homogeneity which may facilitate the implantation of invasive exotic species in the port environment and their dispersal outside the ports [10,17]. In their natural habitat, the distribution of species along the intertidal zone is in part related to the environment stress engendered during emergence at low tide or severe weather conditions. Yet many smooth-surfaced concrete structures do not provide habitats, deprive marine species of shelter against the effects of emersion and reduce their chances of survival. Various studies carried out worldwide have evidenced the interest of investigating the nature of port constructions, whether in terms of the choice of materials, their water-retentive capacity, or the creation of micro-habitats in such a way as to improve the local ecological potential [18, 19, 20]. For example, water retention related to the porosity of the material or the presence of cavities has been described as a factor favouring biodiversity in an urbanised environment [18, 21, 22, 23]. Modification of the surface of the concrete such as by grooved surfaces has enabled an increase in colonisation, the coverage rate and the species richness [19, 24, 25]. The addition of tide pools has also enhanced the specific richness on pre-existing dykes compared to smooth surfaces [26, 27]. This research has led to the development then the commercialisation of so-called 'biologically active' or 'bio-active' concrete of which the intrinsic (formulation) and surface characteristics (porosity, roughness) should facilitate the installation and the development of marine life in urban environment (e.g. [28]).

To our knowledge, no eco-engineering study has been undertaken in southern Brittany in port environment to test such improved concrete. In order to restore the biodiversity in port environment and contribute to limiting the erosion of the biodiversity, we have undertaken for the first-time eco-engineering studies based on so-called 'ecologically active' concrete in the port of Port Haliguen (Quiberon, France). The hypothesis of our study was that the different concretes tested would exhibit a different process of colonisation. To test this hypothesis, we tested various commercial concretes (including one produced by 3D-printing) expected to be more favourable because of their intrinsic characteristics (surface pH, roughness associated with the prefabrication method or with a matrix, porosity and water retention during the tidal cycles). In order to take into account the configuration of port infrastructures, initial design prototypes in the form of panels or slabs of smaller were tested. More specifically, the goals were (1) to ensure that the various concretes tested were biocompatible and suited to immersion inside a port in the South of French Brittany, (2) to monitor the rate of their colonisation by macrofouling at 2 tidal levels (3) to propose specific recommendations in matters of in-port ecological engineering for the Brittany region (France) at different tidal levels.

2. Materials and Methods

2.1. Description of the study site

Port Haliguen (47.4858°, -3.0995°) is a port for pleasure craft and coastal fishing vessels situated at the south-east end of the Quiberon peninsula (Morbihan, France). The tidal range varies by 5.30 m at the time of the spring tides. This study was carried out at two sites in the port, at the northern end of the central concrete pier (tests of colonisation on slabs) and on the opposite dyke (installation of panel prototypes) in such a way as to cover the widest possible intertidal zone while respecting the existing port infrastructures (figures 1 and 2). Thus, to avoid damaging the piles of the central pier, only slabs lighter than the panel prototypes were installed.

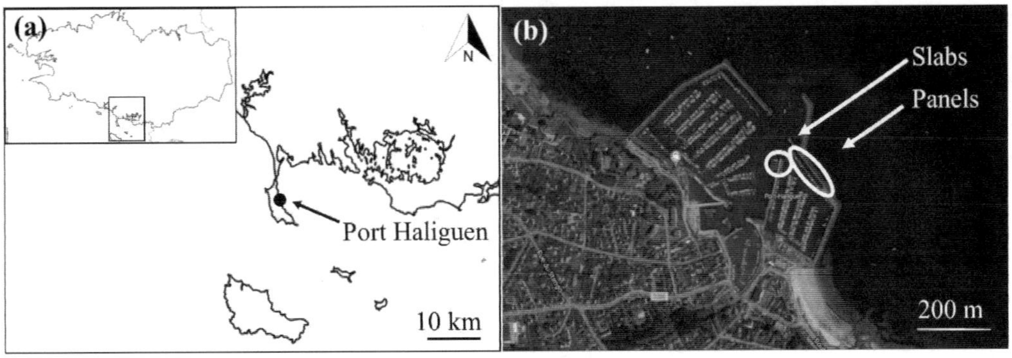

Figure 1. Emplacement of the two experimental sites at Port Haliguen (Morbihan, France).

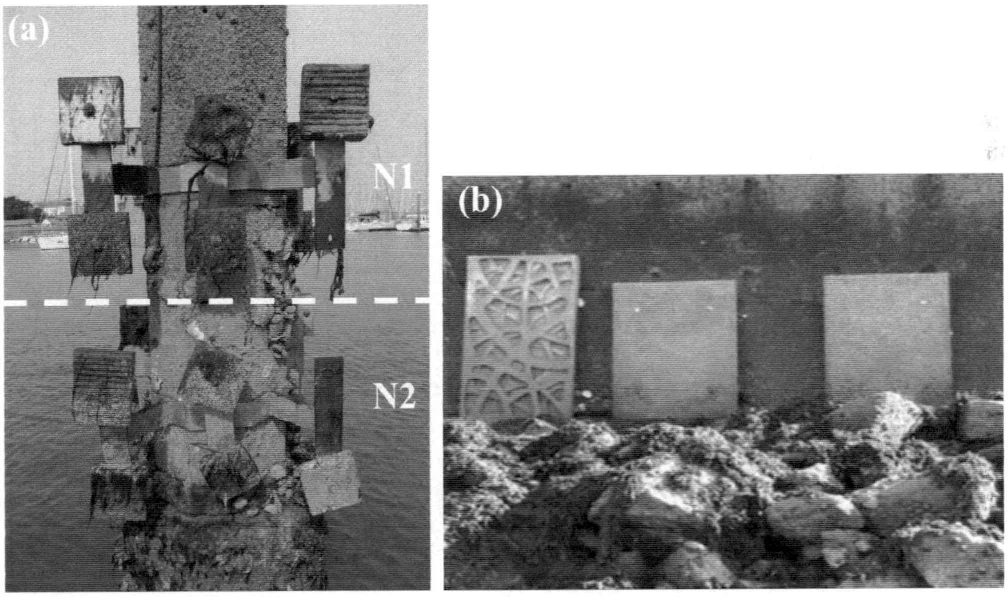

Figure 2. Disposition (a) of the slabs at levels N1 and N2 on the piles of the concrete pier and (b) panels along the dyke at Port Haliguen (Morbihan, France) at level N1.

In parallel, a study of the existing biocenoses at Port Haliguen was undertaken at low tide. Three facies may be observed on the piles of the concrete pier, according to the height of the water and from top to bottom: one Enteromorpha facies of the genus *Ulva* sp. characterising level N0 (+ 5–6 m above the Lowest Astronomical Tide (LAT)); one facies of the barnacle *Chthamalus montagui* and limpets characterising level N1 (+ 3–5 m LAT); this is followed by a facies of *Magallana gigas* and *Mytilus edulis* mixed with *Ulva* spp. green algae and brown algae of the species *Fucus vesiculosus* characterising level N2 (+ 1–3 m LAT). Opposite the pier, the dyke exhibits on the inner wall facing south-west a different specific richness and composition. The dyke is for the most part covered with *Ulva* sp. and, at the lower level, a few young Fucales and associated grazers such as *Littorina saxatilis*, *L. obtusata*, *Patella vulgata*. There are also filter feeders, mainly barnacles *Chthamalus* sp., a few *Austrominius* sp, and a few bivalves *Magallana gigas*.

2.2. Materials and experimental design

Five different formulations of concrete (A, B, C, D and E), judged empirically optimal on the basis of the main conclusions of experimental tests undertaken worldwide and recorded in the bibliography, were used. Thus, in comparison with a smooth concrete traditionally used in maritime engineering, the concretes tested here were optimised in terms of the roughness of the surface, porosity, pH or construction method (table 1). Formulation A aimed to recreate the surface effect traditionally observed in maritime engineering [24, 29]. Formulations B and C judged bioactive with reduced pH, or close to that of sea water were designed to enhance the complexity of the substrate at small scale (< 1 cm), the porosity and the roughness of the surface [19]. Formulation B produced a material that was entirely porous on the whole surface and Formulation C solely on the outside surface over half of its thickness (bi-layered). Formulation D was matrixed in such a way as to increase the roughness of the surface (>1 cm) and with diminished pH, close to that of the sea water. Finally, Formulation E was C3Dmix, non-fibrous mortar designed using CEM I (Parenxlanko) 3D printable (Constructions-3D, Valenciennes, France). Due to the fabrication process, this formulation presented high roughness and grooving (1-2 cm). Sixty slabs (20 x 20 x 5 cm) were fixed vertically at the northern end of the central pier in July 2018 (t_0) on 3 piles spaced of 5 m at two levels of high tide and intermediate tide, N1 and N2, respectively (figure 2). Six replicates per formulation were studied along the central pier. A random positioning of the formulations was used for each treatment.

Table 1. Description of the 5 materials used as slabs

Slab	Formulation	Supplier	Replicates	Location	Installation date
A	CEM III concrete, smooth, non-porous and non-matrixed	LafargeHolcim Ltd, Jona, Switzerland	6 at High Tide and 6 at Mid Tide	Concrete pier	July 2018
B	CEM III concrete, porous with reduced pH	LafargeHolcim Ltd, Jona, Switzerland	6 at High Tide and 6 at Mid Tide	Concrete pier	July 2018
C	CEM III concrete, porous with reduced pH	LafargeHolcim Ltd, Jona, Switzerland	6 at High Tide and 6 at Mid Tide	Concrete pier	July 2018
D	Special "Atlantic" cement	ECOncrete® Tech Ltd, Tel-Aviv, Israel	6 at High Tide and 6 at Mid Tide	Concrete pier	July 2018
E	CEM I	Constructions-3D, Valenciennes, France	6 at High Tide and 6 at Mid Tide	Concrete pier	July 2018

Three types of complex surface panels (P2, P3, P4) were installed in March 2019 every 45 m along the north-east dyke (230 m) at Port Haliguen (figure 2). Panels were positioned vertically against the wall. The formulation of panel P3(163 x 113 x 20 cm) is identical to the formulation of slabs B (table 2). Formulation of panel P4 (163 x 113 x 20 cm) differs from formulation P3 in the use of a granular fraction that is twice as fine, expected to be associated to a higher water retention. P4 was only studied on the prototype panels. The third panel, P2, (180 x 90 x 15 cm), was similar to Formulation D. Since the 3D printing did not allow such dimensions within the project deadline, panels with an equivalent formulation to the type E slabs could not be made and tested. A random positioning of the panels was deployed in triplicate along the dyke.

Table 2. Description of the 3 materials used as panels

Panel	Formulation	Supplier	Replicates	Location	Installation date
P2	Special "Atlantic" cement	(ECOncrete® Tech Ltd, Tel-Aviv, Israel)	3	Dyke	March 2019
P3	CEM III bilayer porous concrete	LafargeHolcim Ltd, Jona, Switzerland	3	Dyke	March 2019
P4	CEM III bilayer porous concrete	LafargeHolcim Ltd, Jona, Switzerland	3	Dyke	March 2019

2.3. Biological monitoring
For each slab, 9 monitoring surveys between September 2018 and August 2019 enabled to assess the coverage rate by the sessile fauna and flora and the composition associated with each formulation at the two levels defined previously. The surveys were carried out on the basis of photographs (60 photos per monitoring survey) by scuba diving during mean high-water neap. The photographs were taken with a PowerShot G12 (Canon®) camera and analysed *a posteriori* in the laboratory using the software ImageJ in order to calculate the coverage rate of the slabs. The area covered by marine organisms on the slabs was calculated as the percentage of the area covered by each taxon in relation to the total surface of the slab.

On each panel, 3 photoquadrats (20 x 20 cm) were done at three different levels: on the upper part, on the middle part and on the lower part of the panel (+3.50-3.7 m LAT). The vertical wall of the existing concrete construction (P1) served as reference with 3 photographs taken at the same three levels at each site. The surveys were carried out monthly from April 2019 to September 2019 during low water in spring tides in order to assess the specific richness and the coverage rate by Fucales. As for panels, photographs were analysed in the lab using ImageJ and covered rate was calculated as previously as the ratio of surfaces.

2.4. Statistical analysis
The data were then reprocessed using the software RStudio 4.0.3. The one factor ANOVA test was used to compare on a monthly basis the mean coverage on each slab. The PERMANOVA test (999 permutations, Euclidian method) associated with a Tukey Post-hoc test allowed comparison of the percent coverage in Fucales of the experimental panels at the lower level of the panels (+3.5-3.7 m LAT).

3. Results
3.1. Monitoring of the slabs
At the upper level N1, the coverage rate on the 5 types of slabs increased constantly to reach 100% by February (figure 3). Between the months of September (t_0+2 months) and October (t_0+3 months), the concretes tested were rapidly colonised by a biofilm to which were added the first pioneer species, mainly Chlorophyta of the genus *Ulva* sp. (figures 4 and A1). At this stage, the coverage rate of the type A slab (smooth concrete) was significantly higher than the other materials (p-value = $0.175\ 10^{-3}$). From January (t_0+6 months), these differences were no longer significant. From February, the coverage rate reached 100% whatever the type of concrete. From May, barnacles of the genus *Austrominius* sp. gradually made their appearance on type A and E slabs. The coverage by the green algae *Ulva* sp was

at a maximum during the summer. Between July and August (after one year of immersion), brown macro-algae of the genus *Fucus* sp. made their appearance on type B, C and D slabs.

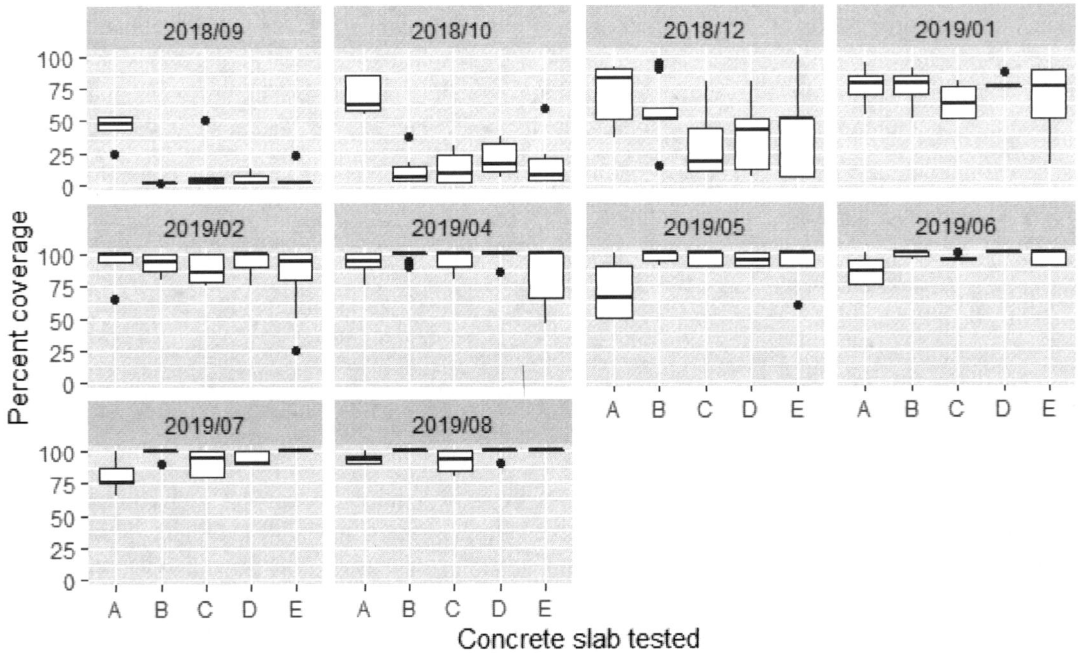

Figure 3. Coverage rate of concrete slabs (A, B, C, D and E) set at level N1 between September 2018 and August 2019.

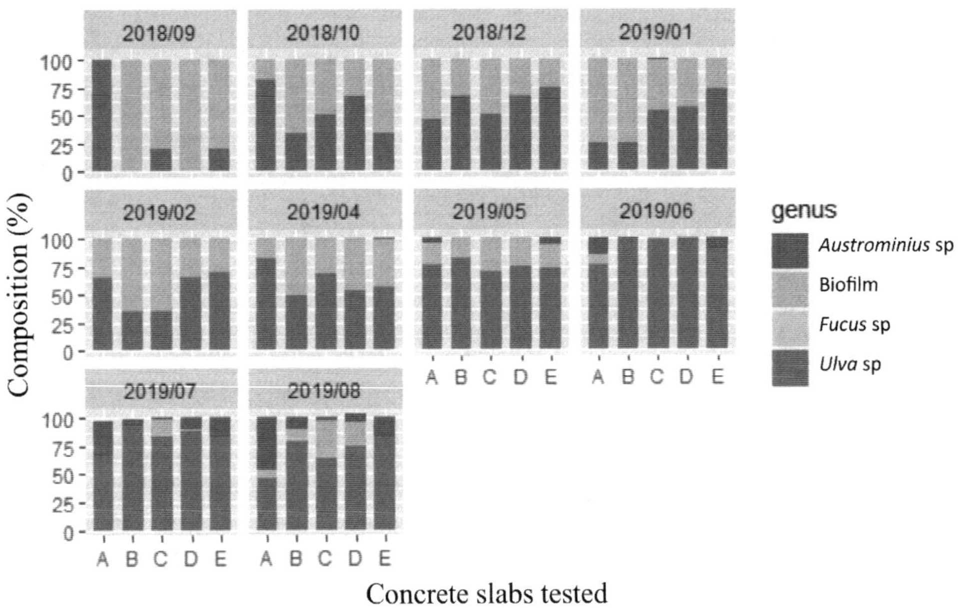

Figure 4. Patterns of change in the floristic and faunal composition on the five types of slab at level N1 over time.

At the intermediate level N2, as soon as September, the type A and D slabs were rapidly colonised by encrusting species and Chlorophyta (figure A1). By October (t_0+3 months), the coverage rate on A, B, C, D and E slabs was already high, ranging from 10 to 75%. As for level N1, the coverage rate of all the slabs reached almost 100% in February (figure 5). As for level N1, the type A and E slabs seemed to favour the presence of barnacles of the genus *Austrominius* sp. These appeared early, from September on the type E material and then in February on the type A material (figure 6). Towards July, all the slabs were covered by these barnacles and still for the most part on the type A and E slabs. During the summer, a few herbivorous gastropods of the genus *Steromphala* sp. or *Patella* sp. made their appearance. Between July and August, after one year of immersion, brown macroalgae of the genus *Fucus* developed on the type C and E slabs.

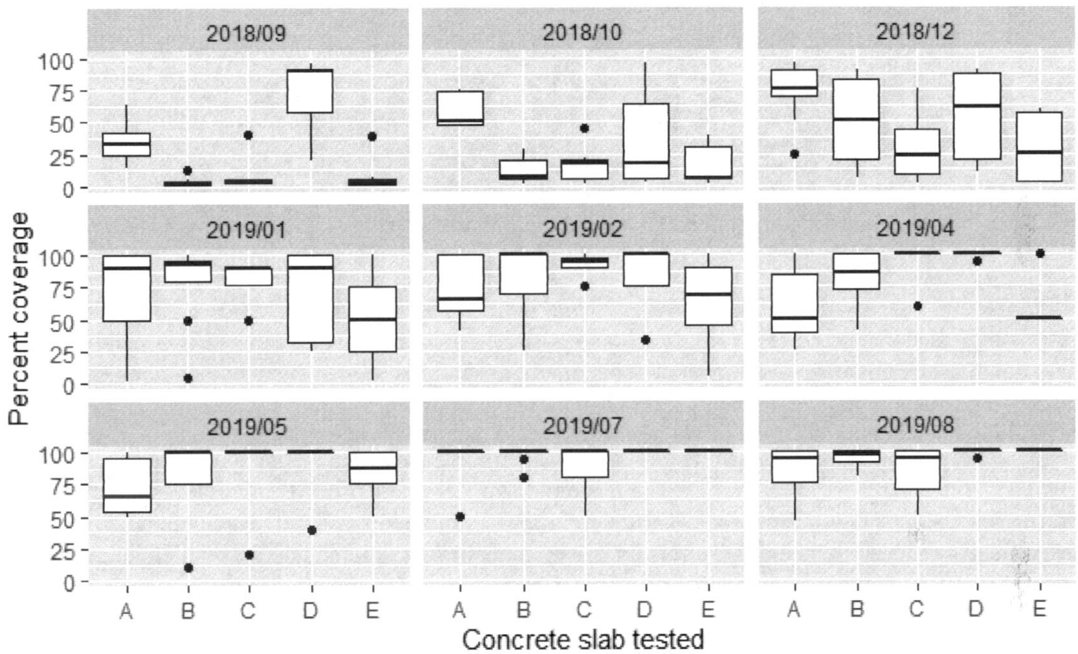

Figure 5. Coverage rate on immerged slabs at level N2 between September 2018 and August 2019.

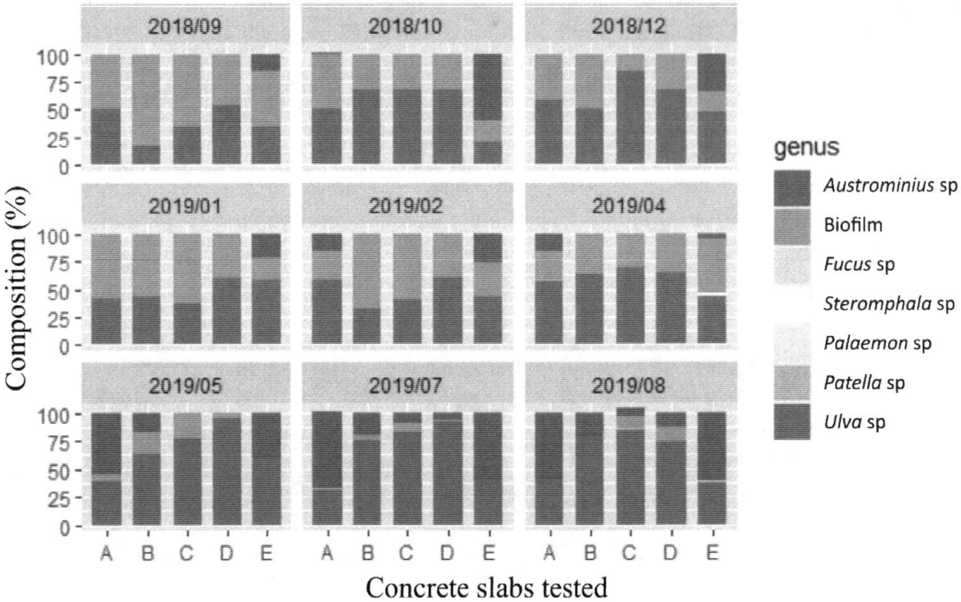

Figure 6. Patterns of change in the floristic and faunal composition on the 5 types of slab at level N2 over time.

3.2. Monitoring of prototype panels

Regarding the complex surface panels, colonisation by marine organisms also varied in function of time and according to the nature of the panels without however reaching the specific richness of the existing structure (i.e. P1) (figure 7). For example, on matrixed panel P2, barnacles of the genus *Austrominius* sp. made a massive appearance in June, accompanied by *Ulva*. On the porous panel P3, the surface was covered from May (t_0+1 month) by a turf of microalgae. A few *Ulva* developed progressively but no barnacle was observed in contrast to panel P2 (figure A2).

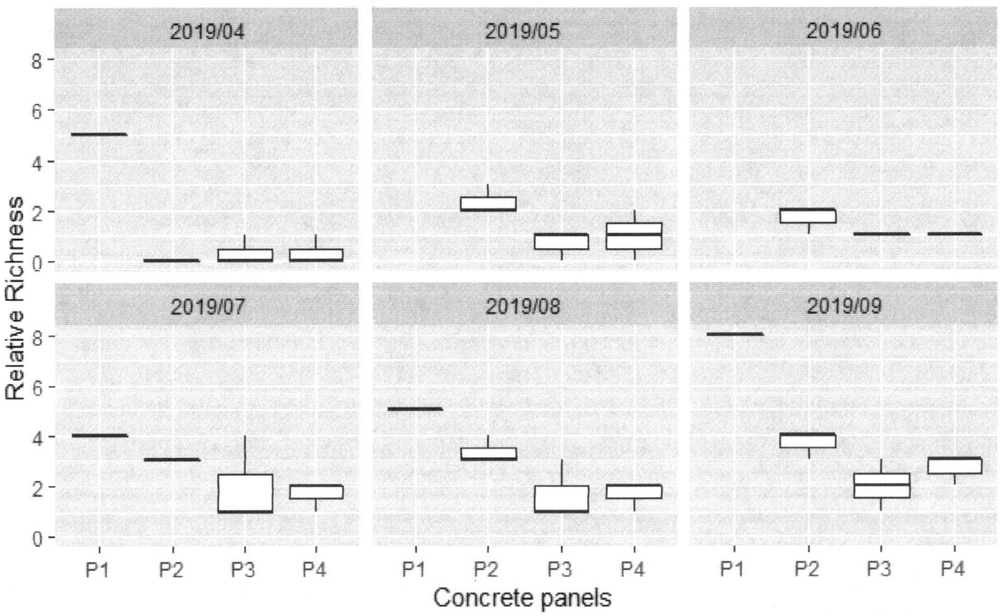

Figure 7. Patterns of change in specific richness on panels in function of time

It is on the lowest level of the panels that the communities are most complex, with notably the appearance of the brown macro-algae *Fucus vesiculosus*. Despite a low number of replicates (n = 9), the panels exhibited a different coverage percentage of Fucales in function of the materials tested (Pseudo – $F_{3,24}$ =75.3, $p < 0.001$). The coverage of Fucales was significantly higher (figure 8 and A2) on the P4 concrete panels (70.3 ± 21.3) after 6 months of study in comparison with the other treatments at level +3.5-3.7 m LAT (p value < 0.001).

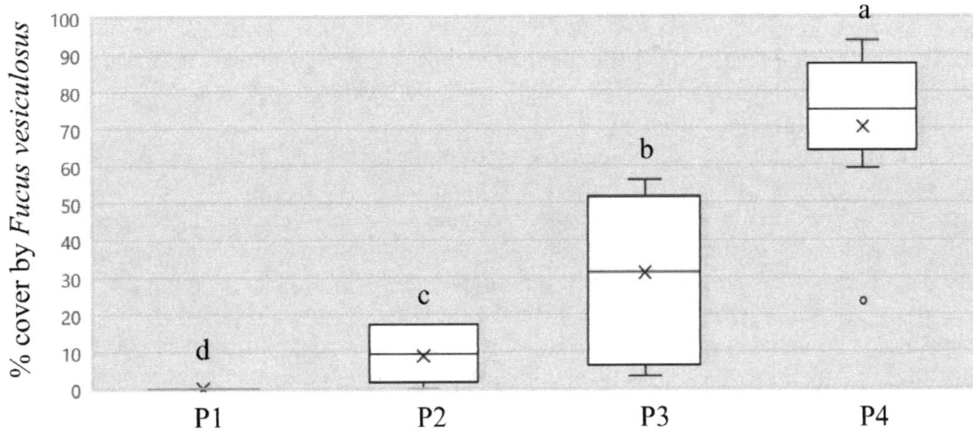

Figure 8. Coverage percentage of Fucales on different panels in September (n = 9) after 6 months of installation at level +3.5-3.7 m LAT. Permanova with Post-hoc Tukey test. Different letters represent significant differences associated with p-value *** $p < 0.001$.

4. Discussion

All the materials and prototypes tested in this study proved to be suited to the functioning of a dynamic port space such as that of Port Haliguen. After one year of experimentation, the constructions were still in place and structurally in good condition. This is notably the case for the 3D printed concrete structures which do not at this stage show structural weaknesses despite the local tidal conditions. The advantages of this new technology are many and might be put to good use in future ecological engineering projects, notably in in-port environment. Using 3D printing, it is today possible to purpose-build more heterogeneous surfaces and to provide ecological niches close to natural conditions, in contrast to traditional constructions [29]. This may for example involve creating cavities and water retention zones within tide pools or gentler slopes in such a way as to limit disruption of the connectivity on the vertical structures [19, 21, 22, 29]. By enhancing colonisation by engineer species, these can in turn influence the recruitment of other species by altering the environmental conditions, providing new habitats and / or supplying other resources and thus performing additional ecological functions [30, 31].

In a general way, at the level of the intermediate and lower midlittoral (N1 and N2, respectively), fouling of the slabs installed and the panels was limited after one year of study mainly to three major taxonomic groups: macro-algae (green and brown), filter-feeders (barnacles) and grazers (Littorina, gibbles, limpets), classically observed on concrete structures along the southern coast of the British Isles [24]. Depending on their intertidal distribution, the different prototypes tested in this study presented a variable kinetic of colonisation and an increasing gradient of specific richness from High Tide to Low Tide level. Port constructions and ecological engineering designs tested worldwide on vertical or steeper slopes indicate furthermore that these intertidal levels remain less diversified in comparison with the lower shore, whether in natural or artificial environment [24, 32-34]. As natural intertidal shores can represent tens of meters in width, artificial vertical structures greatly reduce this area only to few meters and highly limit the settlement of diverse communities.

On the slabs, it is type A (smooth concrete) which unexpectedly appeared to be more rapidly colonised than the four other materials at levels N1 and N2 of immersion, that is above sea level at low water in spring tides, by pioneer species such as the Chlorophyceae; these differences rapidly disappear with time. Contrary to the expected results [25, 28, 35, 36], there were no differences between the slab treatments, regardless of the complexity of the surface, roughness, water retention or intrinsic physical or chemical characteristics of the materials that were supposedly enhanced. For example, in the case of 3D printed concrete, it's the current extruder 'process' that provides the characteristic surface texture of the material. Although its formulation was different from the others, there was also no significant differences at the two tide levels. So, it should be noted that it's generally difficult to disentangle the influence on bioreceptivity of formulation 'chemistry' and 'texture/roughness' of the formulation especially in a such short period of time.

The specific study of the panels shows that in 6 months, the brown macro-algae (*Fucus vesiculosus*) had colonised the different prototypes thus increasing locally the zoning of the macro-algal belt towards the top in comparison with the existing structure. The emergence and grazing pressure by herbivores is considered as the main factor establishing the upper limit for many intertidal algal species. The presence of herbivorous grazers at these levels suggests that this high algal zoning is linked to the concrete formulation and not to the absence of predation [18]. Despite the high variability predicted within the port and a limited number of replicates, one formulation, the porous concrete P4, stood out with mean coverage by brown macro-algae significantly higher than the other treatments. It differs in particular from the second bi-layer porous concrete P3 in having a finer granulometry. Compared to P3, the water retention is expected to be higher limiting the desiccation risk during emergence. This porous concrete could combine the surface characteristics (porosity, water retention, roughness) best suited to the development of this engineer species, *a minima* in terms of the kinetic (first 6 months), by facilitating the fixing of zygotes (dioecious species) and the maintenance of the holdfast of new individuals of *Fucus vesiculosus* during the first six months. However, it's advisable to remain very cautious with regard to these preliminary results. It would not be surprising if, in the longer term, the general patchy recruitment and distribution of grazers may well alter the coverage over time [37].
The brown macro-algae are used locally as bioindicators of the quality of the natural environment by the REBENT monitoring programme and are in severe decline at worldwide, European and national scale [38-40]. Yet they are the engineer ecosystems which directly or indirectly benefit numerous taxa of fisheries interest such as fishes and crustaceans [41-43]. This study has shown that in-port ecological engineering solutions could specifically enhance the colonisation of existing and future constructions by brown macro-algae and potentially the associated ecological functions. Other species, such as the barnacles, are also described in the literature as engineer species whose presence (even after their death) might influence the structure of the communities that succeed each other over time [44]. The roughness of the surface of the concretes would here to be one of the factors conditioning the larval recruitment [25].

On the basis of these results, several recommendations might be offered for future studies in southern Brittany:

- achieve local rigorous port communities diagnosis before any ecological engineering actions
- slower the slope of vertical infrastructures such as dykes to extend the distribution and improve the diversity of intertidal communities
- encourage the use of porous concretes with fine grain and controlled granulometry in order to maximise water retention and promote intertidal communities along port infrastructures and especially engineer species such as macroalgae.

- develop 3D printing porous concretes with fine grain and controlled granulometry, water retention areas, allowing monitoring of the marine biofilm and the macro-fouling
- develop eco-design solutions for the whole of the port coastal fringe, from the midlittoral to the upper infralittoral
- particular attention should also be paid to the development of potentially invasive exotic species. In the course of our study on the monitoring of the slabs, materials A and E seemed to be more favourable for the settlement of barnacles of the genus *Austrominius*, probably *A. modestus*, an invasive species well established in southern Brittany. Material A differs from other concretes mainly by its smoothness. Among the different prototype panels tested, it was also the smoothest, that is the matrixed panels (P2), which were massively colonised by the invasive species *Austrominius*.
- at worldwide and at a local scale, the interest of certain ecological engineering approaches such as, for example, reducing the pH of the surface of concretes by using CEM III type cement remains to be confirmed according to certain authors [45]. Given the difficulties of interpretation in ecological engineering pilot projects like our study carried out over short periods, it is recommended today to extend and duplicate this project typology both temporally and spatially in order to confirm or not the first results observed before introducing effective and properly supported restoration and ecological engineering policies [46].

5. Conclusion

This study has enabled to undertake for the first time in Brittany (France) a novel experimental pilot project focused on the development of the potential of ecological engineering for intra-port structures in intertidal zone. It was also the first study undertaken in this region on the potential for colonisation of the surface of bilayer porous concrete and concrete produced by 3D printing. The adopted strategy of high frequency monitoring enabled us to progress in one year in our understanding of the mechanisms of colonisation that could be observed, *a minima* locally, at Port Haliguen. Monitoring of the experimental slabs and panels installed at the midlittoral level showed that certain concrete formulations could play an essential role in increasing the richness of the in-port fauna and flora communities, but these results show high variability depending on the method used (slabs vs panels) and the tidal level studied. This project confirms the difficulty of interpretation of pilot experiments of this kind monitored on a short-term basis, while highlighting the interest of working to enhance our knowledge regarding the environmental integration of existing and future port constructions and the development of their ecological functioning. This work needs therefore to be extended and duplicated in order to confirm the trends observed in time and space (e.g for Fucales) with the aim of achieving a definition of quantifiable goals and offering guidelines that might contribute to the restoration of habitats and ecological functions (reproduction, hatchery, nursery, connectivity) along urbanised coasts under pressure in Brittany.

Acknowledgments

The authors are grateful to Office Français pour la Biodiversité and to Compagnie des Ports du Morbihan and in particular the Capitainerie de Port-Haliguen for financial support and for hosting this experiment on their infrastructure. The authors thank Eiffage, Lafarge Holcim, ECOncrete and Construction 3D for providing concrete prototypes.

Appendices

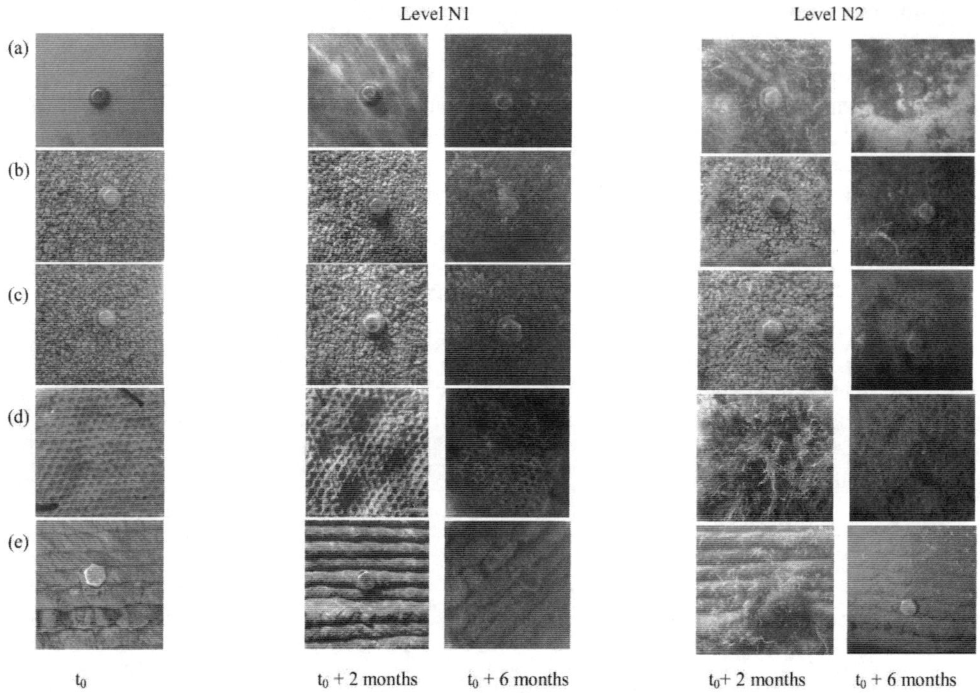

Figure A1. Kinetic of colonisation by *Ulva* spp. on slabs of type (a) A, (b) B, (c) C, (d) D and (e) E set at level N1 (left) and N2 (right) at t_0, t_0+2 months and t_0+6 months.

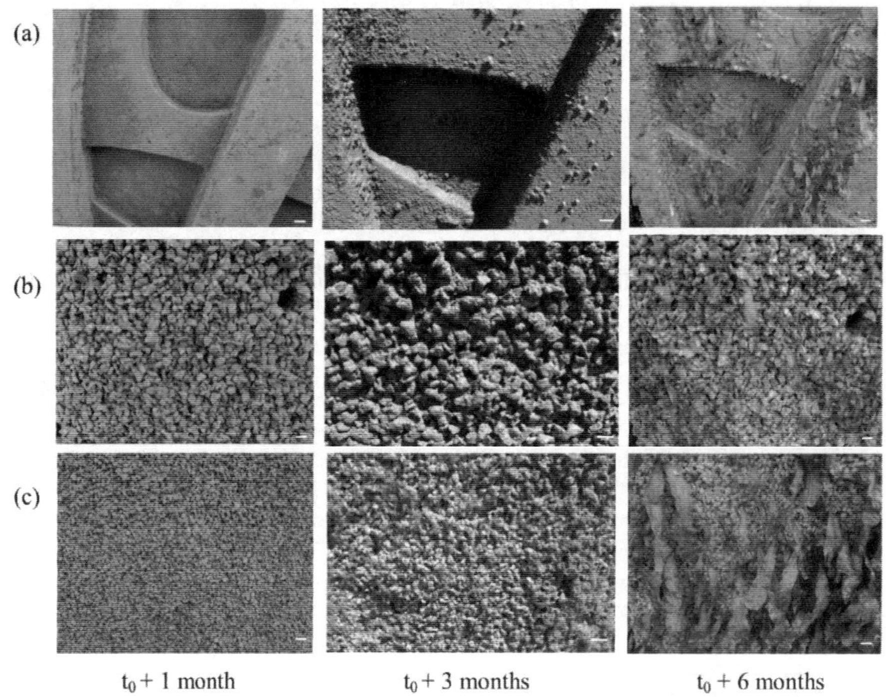

Figure A2. Kinetic of colonisation by Fucales on panels (a) P2, (b) P3, (c) P4 set at +3.5-3.7 m LAT at t_0 + 1 month (April), t_0 + 3 months (June) and t_0 + 6 months (September) (scale bar: 1 cm).

References

[1] Firth L B, Knights A M, Bridger D, Evans A J, Mieszkowska N, Moore P, O'Connor N E, Sheehan E, Thompson R C and Hawkins J 2016 *Oceanogr. Mar. Biol. Annu. Rev.* **54** 189–262
[2] Bishop M J, Mayer-Pinto M, Airoldi L, Firth L B, Morris R L, Loke L H, Hawkins S J, Naylor LA, Coleman R A, Chee S Y and Dafforn K A 2017 *J. Exp. Mar. Biol. Ecol.* **492** 7–30
[3] Nogales B, Lanfranconi M P, Piña-Villalonga J M and Bosch R 2011 *FEMS Microbiol. Rev.* **35** 275–98
[4] Rivero N K, Dafforn K A, Coleman M A and Johnston E L 2013 *Biofouling* **29** 803–15
[5] Momota K and Hosokawa S 2021 *Sci. Rep.* **11** 4028
[6] Di Franco E, Pierson P, Di Iorio L, Calò A, Cottalorda J M, Derijard B, Di Franco A, Galvé A, Guibbolini M, Lebrun J *et al.* 2020 *Mar. Pollut. Bull.* **159** 111450
[7] Davies T W, McKee D, Fishwick J, Tidau S and Smyth T 2020 *Sci. Rep.* **10** 12545
[8] Costello M J and Chaudhary C 2017 *Current Biology* **27** R511–R527
[9] Glasby T M and Connell S D 1999 *Ambio* **28** 595–8
[10] Bulleri F and Airoldi L 2005 *J. Appl. Ecol.* **42** 1063–72
[11] Bulleri F and Chapman M G 2010 *J. Appl. Ecol.* **47** 26–35
[12] Jacob C, Buffard A, Pioch S and Thorin S 2018 *Ecol. Eng.* **120** 585–94
[13] Tan E L Y, Mayer-Pinto M, Johnston E L and Dafforn K A 2015 *Front. Microbiol.* **6** 1276
[14] Chapman M G 2006 *J. Molluscan Stud.* **72** 247–57
[15] Hadfield M G 2011 *Annu. Rev. Mar. Sci.* **3** 453–70
[16] Salta M, Wharton J A, Blache Y, Stokes K R and Briand J-F 2013 *Environ. Microbiol.* **15** 2879–93
[17] Glasby T M, Connell S D, Holloway M G, Hewitt C L 2007 *Mar. Biol.* 151 887–95
[18] Chapman M G and Blockley D J 2009 *Oecologia* **161** 625–35
[19] Firth L B, Thompson R C, Bohn K, Abbiati M, Airoldi L, Bouma T J, Bozzeda F, Ceccherelli V U, Colangelo M A, Evans A *et al.* 2014 *Coast. Eng.* **87** 122–35
[20] Natanzi A S, Thompson B J, Brooks P R, Crowe T P, McNally C 2021 *Ecol. Eng.* **159** 106104
[21] Martins G M, Thompson R C, Neto A I, Hawkins S J and Jenkins S R 2010 *Biol. Conserv.* **143** 203–11
[22] Evans A J, Firth L B, Hawkins S J, Morris E S, Goudge H and Moore P J 2016 *Mar. Freshwater Res.* **67** 123–30
[23] Hall A E, Herbert R J H, Britton J R, Boyd I M and George N C 2019 *Front. Mar. Sci.* **6** 456
[24] Moschella P S, Abbiati M, Aberg P, Airoldi L, Anderson J M, Bacchiocchi F, Bulleri F, Dinesen G E, Frost M, Gacia E, Granhag L, Jonsson P R, Satta M P, Sundelöf A, Thompson R C and Hawkins S J 2005 *Coast. Eng.* **52** 1053–71
[25] Coombes M A, La Marca E C, Naylor L A and Thompson R C 2015 *Ecol. Eng.* **77** 314–23
[26] Firth L B, Schofield M, White F J, Skov M W and Hawkins S J 2014 *Mar. Environ. Res.* **102** 122–30
[27] Browne M A and Chapman M G 2014 *Mar. Ecol. Prog. Ser.* **497** 119–29
[28] Perkol-Finkel S and Sella I 2014 *Proc. From Sea to Shore–Meeting the Challenges of the Sea: (Coasts, Marine Structures and Breakwaters 2013)* (London: ICE publishing) pp 1139–49
[29] Chapman M G 2003 *Mar. Ecol. Prog. Ser.* **264** 21–29
[30] Jones C G, Lawton J H and Shachak M 1994 *Ecosystem Management* (New-York: Springer) pp 130–47
[31] Jones C G, Lawton J H and Shachak M 1997 *Ecology* **78** 1946–57
[32] Browne M A and Chapman M G 2011 *Environ. Sci. Technol.* **45** 8204–7
[33] Chapman M G and Underwood A J 2011 *J. Exp. Mar. Biol. Ecol.* **400** 302–13
[34] Firth L B, Mieszkowska N, Thompson R C, Hawkins S J 2013 *Environ. Sci.: Process. Impacts* **15** 1665–70

[35] Lukens R R and Selberg C 2004 *Guidelines for Marine Artificial Reef Materials* (2nd Ed.) p 205
[36] Loke L H L and Todd P A 2016 *Ecology* **97** 383393
[37] Hawkins S J and Hartnoll R G 1985 *Mar. Ecol. Prog. Ser.* **20** 265–271
[38] Harley C D G, Anderson K M, Demes K W, Jorve J P, Kordas R L, Coyle T A and Graham M H 2012 *J. Phycol.* **48** 1064–78
[39] Raybaud V, Beaugrand G, Goberville E, Delebecq G, Destombe C, Valero M, Davoult D, Morin P and Gevaert F 2013 *PloS one* **8** e66044
[40] Méléder V, Populus J, Guillaumont B, Perrot T and Mouquet P 2010 *Mar. Biol.* **157** 1525–41
[41] Lippert H, Iken K, Rachor E and Wiencke C 2001 *Polar Biol.* **24** 512–22
[42] Teagle H, Hawkins S J, Moore P J and Smale D A 2017 *J. Exp. Mar. Biol. Ecol.* **492** 81–98
[43] Kraufvelin P, Pekcan-Hekim Z, Bergström U, Florin A-B, Lehikoinen A, Mattila J, Arula T, Briekmane L, Brown E J, Celmer Z *et al.* 2018 *Estuar. Coast. Shelf Sci.* **204** 14–30
[44] Boulding E G and Harper F M 1998 Aspects of Littorinid Biology. Developments in Hydrobiology vol 133 ed. R M O'Riordan, G M Burnell *et al.* (Dordrecht: Springer) pp 105–114
[45] Hsiung A R, Tan W T, Loke L H, Firth L B, Heery E C, Ducker J, Clark V, Pek Y S, Birch W R, Ang A C F *et al.* 2020 *Mar. Ecol. Prog. Ser.* **656** 193–205
[46] Firth L B, Airoldi L, Bulleri F, Challinor S, Chee S-Y, Evans A J, Hanley M E, Knights A M, O'Shaughnessy K, Thompson R C and Hawkins S J 2020 *J. Appl. Ecol.* **57** 1762

Development of laboratory tests to measure the growth kinetic of micro-algae of concrete in marine environment and comparison to in-situ experiments

V Morin[1], S Duchand[1], I Dubois-Brugger[1], M Perrot[2], M Lapinski[2] and S Vacher[3]

[1] LafargeHolcim Innovation Center, Saint Quentin-Fallavier, France
[2] Seaboost, Montpellier, France
[3] Conidia, Quincieux, France

Abstract. Climate change associated with population growth drive the building sector's innovation in developing new ecological and sustainable concrete solutions. Regarding marine environment the rise of sea level is a main concern for several big cities across the world. New sea defense infrastructure around cities is expected to rise in the coming years. This will not be possible and accepted without any ecological compensation and thus by developing new ecological concrete range. The development of Bio-active concrete is one part of the solution: in addition to their conventional structural performance, these concrete promote marine bio-fouling and biodiversity. To reach this objective, a laboratory test was developed to screen different concrete surfaces and compositions. The bio-receptivity capacity of a concrete surface is measured by the amount of micro-algae on the concrete surface after four weeks test. This quick lab test supported the development of a new range of concretes called Bi-Layer porous concrete made up of one structural layer and one highly porous bio-active layer. These concrete were implemented as quay walls and reefs in Mediterranean and Atlantic sea. Very promising results were obtained after 18 months of immersion regarding bio-fouling activity and micro-algae diversity on the concrete confirming the results given by the laboratory test.

Authors	E-mail	Postal address
V. Morin	vincent.morin@lafargeholcim.com	95 Rue du Montmurier, 38291 Saint-Quentin-Fallavier, France
I. Dubois-Brugger	isabelle.dubois-brugger@lafargeholcim.com	95 Rue du Montmurier, 38291 Saint-Quentin-Fallavier, France
S. Duchand	sylvain.duchand@lafargeholcim.com	95 Rue du Montmurier, 38291 Saint-Quentin-Fallavier, France
S. Vacher	sebastien.vacher@conidia.fr	Parc d'activité en Chuel, Route de Chasselay, 69650 Quincieux, France
M. Perrot	Martin.PERROT@seaboost.fr	889, rue de la Vieille Poste – CS 89017 34965 Montpellier Cedex 2
M. Lapinski	Matthieu.LAPINSKI@seaboost.fr	889, rue de la Vieille Poste – CS 89017 34965 Montpellier Cedex 2

1. Introduction

The marine biodiversity plays a vital role in maintaining the functionality and productivity of marine ecosystems: one of the most important functions of marine ecosystems is the production of organic compounds from sunlight and nutrients (primary productivity) performed for instance by small micro-organisms called phytoplankton. These organisms responsible for primary production form the base of the food chain, used by all life in ocean and at the end by human. These marine ecosystems therefore provide a wide range of goods and services to the human as food, active pharmacological substances, solid raw material, climate regulation, leisure activities... Marine biodiversity thus sustains the functioning of marine ecosystems and the services that it provides. Today, marine biodiversity is declining and this decline becomes more and more critical in coastal areas that are known to support the richest and most diverse marine ecosystems and particularly prone to human activities. Currently, 60% of human population lives within 100 km distance from the sea front and, most of big cities with more than 15 million of inhabitants are located close to the coastal area. So there is an obvious need to develop solutions that will lead to proper conservation and management of marine species [1]. In construction industry, there is a new tendency to take into account ecological aspects in the development a new marine infrastructure [2]. This new market related to "eco-engineering" aims at developing technical infrastructure that will have a positive impact on the marine environment [3-7]. There is an important challenge in designing concrete that fulfill normative requirements in term of mechanical and durability properties and that provide higher bio-receptivity in marine environment [8]. One starting approach in eco-engineering is the design of concrete artificial reefs [9] and a recent study [10] brought renewed interest in these marine elements: this work showed that artificial reefs are the place of a new biomass production and thus increase efficiently fish biomass and cannot be seen only as attractive elements that depopulate natural marine habitats. Up to now, the main development in reef was focused on its design and on its capacity to supply habitat for different fish and animal size [11]. More recent studies showed that surface bioreceptivity of concrete can be obtained and enhanced by slight modification of concrete composition and surface texture for instance: this was confirmed by laboratory studies [12] but also in-situ measurements [13]. The development of a laboratory test for the bio-receptivity of concrete is crucial: this allows to screen a very large range of concrete composition and surface texture in controlled conditions without any extra cost required to carry out colonization follow up in marine environment. In a recent study Hayek [14] presented and verified a novel experimental laboratory approach which can be used to evaluate the bacterial colonization (bioreceptivity and biofilm formation) of cementitious materials in marine environment. The objective of this study is to focus on the development of a laboratory bio-receptivity test to measure the bioactive capacity of a concrete surface by micro-algae and to support the development of a new range of "green" concrete.

2. Materials and Methods

2.1. Concrete mix-design samples

The concrete samples are designed with one type of crushed calcareous aggregate with a narrow particle size distribution i.e. 1.6 – 3 mm from Cassis quarry (France). Blended cement is used: a "CEM III/B 42.5 LH-SR (from Lafarge La Malle Plant) that contains 65% of slag and 35 % of "pure" Portland cement in order to control the pH of concrete materials in marine environment.

The Bi-layer porous concrete samples are formulated in a very different manner that conventional concrete due to the fact that they contain only one type of aggregate and no fine particles and that the volume of cement paste (volume of cement and water) cannot completely fill the empty volume or porosity around this aggregate. Therefore, when fresh concrete is casted, we use a vibrating table to force the cement paste to move downward and to form a dense concrete layer at the bottom of the mold. As a result, we get a porous layer on the top of the concrete made of aggregate coated by a thin layer of cement paste like it is done in the production of a pervious concrete. The volume of cement is adjusted to get around 80% of the total concrete thickness made of dense layer and 20% made of porous layer (Figure 1). In a nutshell, this concrete mix design enables to get two different types of microstructure (dense and highly porous) gathered in one concrete, also called bi-layer concrete, and made also in one time. So, the upper layer concrete layer is based on an highly porous and permeable concrete with around 30% of porosity. The porosity of the bottom dense layer is around 12% which in line with conventional concrete porosity.

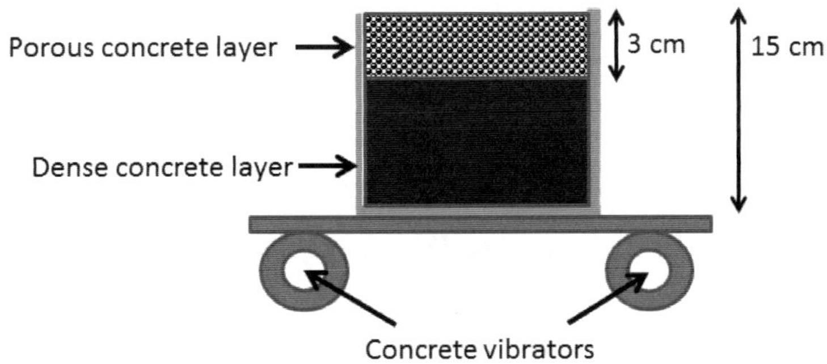

Figure 1. Bi-layer concrete casted in a 15 cm cubic mould placed on a vibrating table

A superplasticizer is added in the concrete mix design to get good initial workability. The type and the amount of superplasticizer is primordial to formulate this Bi-layer porous concrete: it is necessary to get the right fluidity and viscosity of cement paste that will lead to a dense layer formation at the bottom of the mould and a porous upper layer with uniform coating of aggregate by cement paste. More details about the manufacturing process and concrete performance can be found in the patent WO2018042240 [15]
For the bio-receptivity lab test, the concrete samples are casted in a mold with dimension 10x10x3 cm. One day after mixing, the sample is removed from the mold and placed in a sealed bag that maintains a relative humidity close to 100%.

2.2. Experimental set-up
The laboratory test developed to assess the bio-receptivity of materials consists of the colonization evaluation of concrete surface by micro-algae in enriched artificial seawater medium. At age of 28 days, two concrete plates with dimension 10x10x3cm are placed vertically in each tank and the medium is stirred continuously. This stirring is used to maintain liquid solution aeration and to avoid micro-algae sedimentation at the bottom of the tank. For each concrete reference, two plate samples are produced and theses plates are distributed in two tanks at a different location: following this configuration, all concrete reference experience the same hydrodynamic conditions. In order to avoid pH perturbation in the tank from one sample to the other, the two concrete samples placed in the tank have to be produced with the same cement.

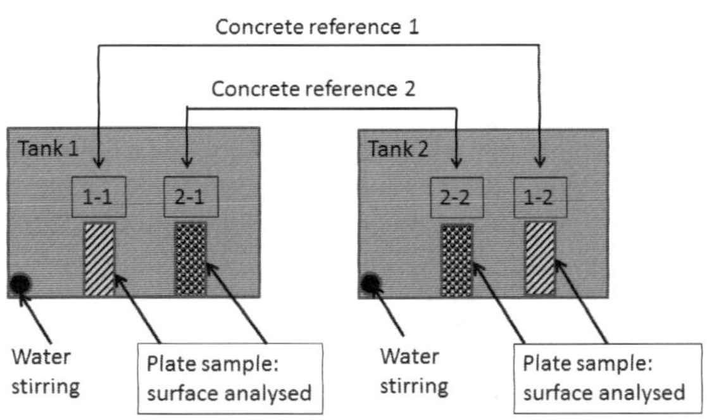

Figure 2. Concrete samples dispatch and positioning in the tank

In the following Table 1 are listed the main parameters of the test:

Table 1. Main parameters of the laboratory bio-receptivity test

Medium			Concrete sample		Vol. ratio: medium/samples (in each tank)	Pump stirring
K medium in artificial seawater	Volume 10 L.	Temperature 20°C	Dimension 10x10x3cm	Volume 0.3 L.	~16	50-100 L/hour
Micro-algae		**Lights**			**Test duration**	
Entomoneis species	[C]: 10^4 /ml	Artificial light: 2500 Lux	Simulation of day/night alternation (12h/12h)		28 days with a complete medium regeneration at T0+14 days	

The micro-alga *Entomoneis* was chosen as standard species in our test and its selection was based on literature review and on its ability to growth on concrete surface [16]. These micro-algae are grown in Roux flask in a specific aqueous solution called K medium based on sterilized and filtered seawater. However in the test, it is possible to use a mix of different microalgae in a same tank.

Table 2. K medium composition used in micro algae cultivation

Reference	volume
Sterilized and filtered seawater	1 L
Sodium Nitrate ($NaNO_3$) at 75 g/l	1 ml
Ammonium Chloride (NH_4Cl) at 2.68 g/l	1 ml
Na_2-Glycérophosphate ($C_3H_7O_6PNa_2$), 5-6H_2O at 3.06 g/l	1 ml
Tri-base (pH 7.2) at 121,1 g/l	1 ml
K trace Metal solution	1 ml
Vitamin solution F/2	0,1 ml

2.3. Micro-algae colonization measurement

The concentration of micro-algae on the concrete surface is measured via a non-destructive method with BentoTorch equipment. BenthoTorch is a portable field instrument and an in vivo fluorometer designed to quantify algal biomass. The amount of red light fluorescence arising from excitation at different wavelengths gives a quantitative estimate of the algal density (expressed in µg of chlorophyll/cm^2) and its classification. The BenthoTorch is precalibrated for the most prominent algal classes to be found in the field and no sample preparation is required. In order to get more reliable results, ten measurements are realized on each concrete surface at determined maturity time: T0+7, T0+14, T0+21 and T0+28 days. At each maturity time, the sample is taken out of the tank: the surface is washed with artificial seawater before to proceed with the analysis using Bentotorch and then the sample is put back in its place without any perturbation of its surface colonization.

2.4. Description of the REXCOR's and ANTICELTO pilot projects
In addition to the laboratory bio-receptivity tests, the colonization of experimental slabs and panels of porous Bi-layer porous concrete were immersed and monitored as part of two pilots for ecological port engineering projects with complementary objectives carried by Seaboost. Indeed, if laboratory tests need to be standardized, positive in-vitro results must be confirmed by field experiments. Two experimental studies are introduced in this article to illustrate the approach and positive results confirming the interest of working on the bio-receptivity of concrete.

The first project, REXCOR, is an ecological restoration project carried out since 2018 in the Calanques National Park (Mediterranean, France). 12 Ragues © type artificial reefs were made using porous Bi-layer porous concrete technology and submerged in 4 villages between – 21 meters and – 24 meters deep. In addition, the benthic colonization for 3 formulations of Bi-layer porous concrete materials has been studied for 2 years according to various environmental parameters known to influence ecological successions (orientation to light, distance to a pressure source, here a discharge from wastewater treatment plants.). 18 slabs (20 x 20 cm) of 3 different Bi-layer porous concrete formulations (A, B and C) were prefabricated and immersed in two stations between – 21 meters and - 24 meters deep and installed in 3 orientations (vertical, horizontal exposed to light, horizontal not exposed to light). The monitoring was carried out using fixed photoquadrats during 4 campaigns and the results were compared with a "reference" ordinary concrete (D) with a CCV type smooth surface.

The second project, ANTICELTO, is a port ecological engineering project carried out since 2018 in the port of Port-Haliguen (Atlantic, France). The benthic colonization for 2 formulations of Bi-layer porous concrete materials (A and B) was studied in the intertidal zone along a vertical structure during 1 year at 3 stations located on the North-East dike of Port-Haliguen. 3 panels of each Bi-layer porous concrete material formulations (113 x 163 x 20 cm) were distributed among the three stations. The monitoring was carried out during 9 campaigns from September 2018 to August 2019 using fixed photoquadrats (20 x 20 cm) carried out at 3 different tidal levels N1: + 5-4.80m CM, N2: + 4.90-5.10m CM, N3: + 3.50-3.70m CM. The results were compared with a "reference" ordinary concrete (C) constituting the existing structure over 50 years old at these same levels with an identical number of quadrats.

3. Results

3.1. Bio-receptivity lab test results
The following figure 3 present microalgae concentration on concrete surface as a function of time for the different concrete samples. The first sample is the Bi-Layer porous concrete with a highly porous and permeable surface concrete (called porous surface) and made with CEM III/B cement and aggregate 1.6-3 mm. The second concrete (called dense surface) is based on only the dense layer and made with the same CEM III/B cement and aggregate 1.6-3 mm. Surface pH of both samples was checked before immersion: the operating method is not described in this study but the pH value was measured at around 12.
Regarding colonization process, we observe a clear increase of microalgae concentration on the highly porous surface that reach around 1.5 µg / cm^2 at 28 days. At the opposite, the micro-algae concentration remains low on the dense concrete surface event after 28 days of immersion. These results were repeated several times and confirm the higher bio-receptivity of the porous, permeable concrete.

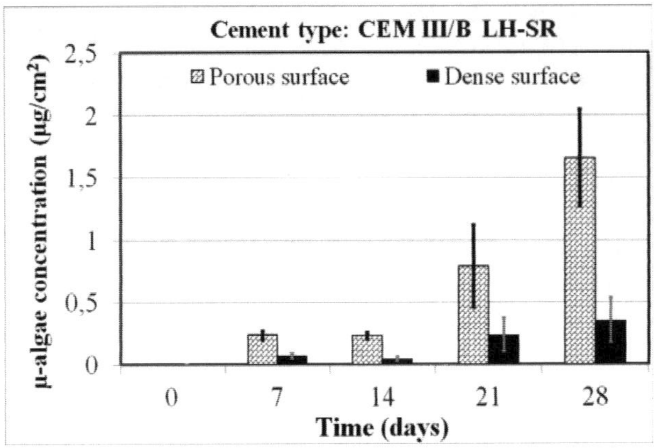

Figure 3. Kinetic of surface colonization between a highly porous and dense concrete

These results showed that a microstructure that presents surface roughness and that is highly porous and permeable is a more powerful support for micro-algae colonization. This porous structure should be able to filter and trap micro-organisms that will settle and grow in the concrete porosity.

3.2. REXCOR's project results or/and ANTICELTO project results

The REXCOR project confirmed the possibility of manufacturing Ragues © type artificial reefs (10t for 3m in height) entirely in Bi-layer porous concrete and without using reinforcements thanks to Seaboost's patented design.

Figure 4. Rague Reef© (Seaboost)

Figure 5. Bi-Layer Concrete (LafargeHolcim) used in Rague Reef© (Seaboost)

More than 60 species of vagile macrofauna (fish, crustaceans, cephalopods) have been observed in the reef villages, 60% of which in the cavities of the Ragues © reefs. Note the regular presence of threatened species such as the groupers Epinephelus marginatus and Epinephelus caninus.

Regarding the monitoring of experimental slabs, the results obtained confirm the results observed in the laboratory test for Bi-layer porous concrete materials in this biogeographical context for pioneer species. Corallinacea, encrusting red algae pioneers in the coralligenous, one of the richest ecosystems in the Mediterranean, developed more rapidly on the three Bi-layer porous concrete materials called "A-B-C" compared to the "reference" ordinary dense concrete studied called 'D" (see Figure 6). Additional monitoring is currently underway in order to study the colonization processes of later species such as bryozarians, sponges and gorgonians in order to confirm the first positive trends observed.

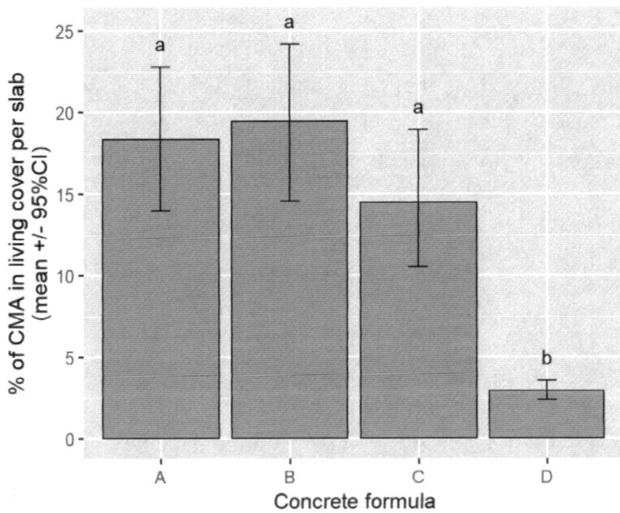

Figure 6. Average Corallinacea coverage per formulation (A-B-C: Bi-Layer porous concrete / D: "reference" ordinary concrete) (n = 72)

Regarding ANTICELTO project, the monitoring in the intertidal zone of experimental panels made with Bi-Layer Concrete, confirms the results observed in the laboratory in this other biogeographic context. The Figure 7 provides the variance data of the observed results (mean, median and quartiles). Two concretes named A and B are based on Bi-Layer Concrete technology but with different aggregate size (1,6 – 3 mm and 4-6 mm respectively). The concrete C is a reference or conventional dense concrete as described before in the document.

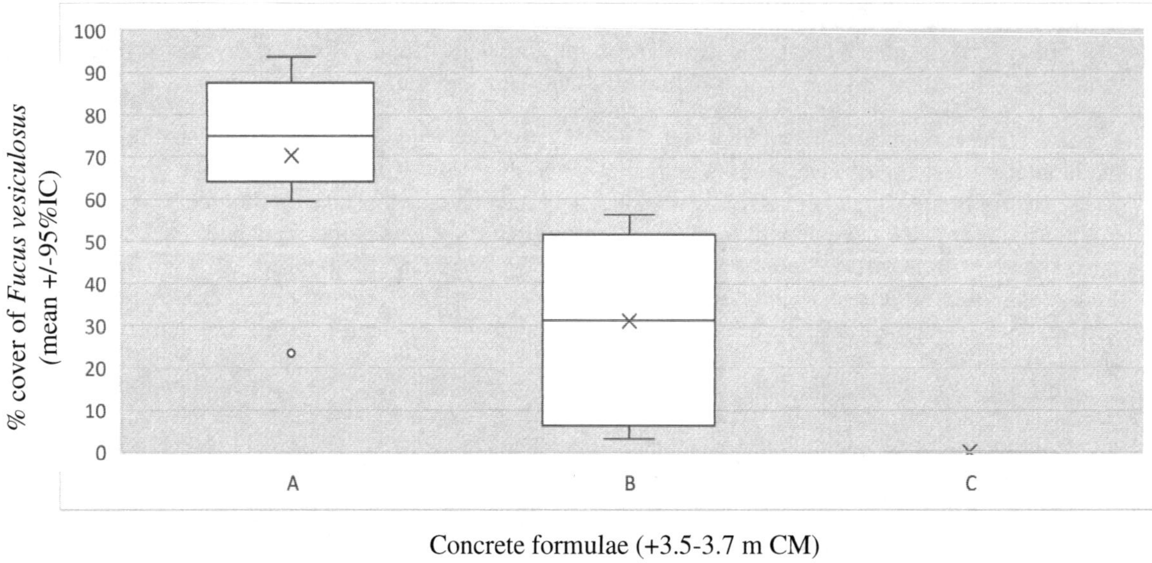

Figure 7. Average coverage of Fucales (Fucus vesiculosus) by formulation (A-B: Bi-layer porous concrete / C: "reference" ordinary concrete) (n = 9) at the lowest level studied (N3: + 3.50-3.70m CM) after 6 months of colonization. Formulation A: Aggregates size from 1.6mm to 3mm. Formulation B: Aggregates size from 4mm to 6mm.

For the Bi-layer porous concrete A, the percentage of coverage is on average 70% with maximum greater than 90% and minimum around 60%. Contrary, for the "reference" ordinary dense concrete C, the result is "zero". In other words, there was no fucale (Fucus vesiculosus) on the "reference" ordinary dense concrete and there was a little on the bottom of panel B (big variability anyway) and a lot on the bottom of panel A (see photos below).
Brown macroalgae are tending to regress in Brittany, but they perform various major ecological functions both when they are submerged (nursery, shelters) and at low tide (temporary shelter against desiccation, feeding the avifauna). Therefore, the results observed here are very positive for this local marine environment.
Complementary monitoring is currently underway to study whether these trends fade over time or, on the contrary, have a real effect on this taxonomic group and on any associated taxa (eg: limpets, periwinkles).

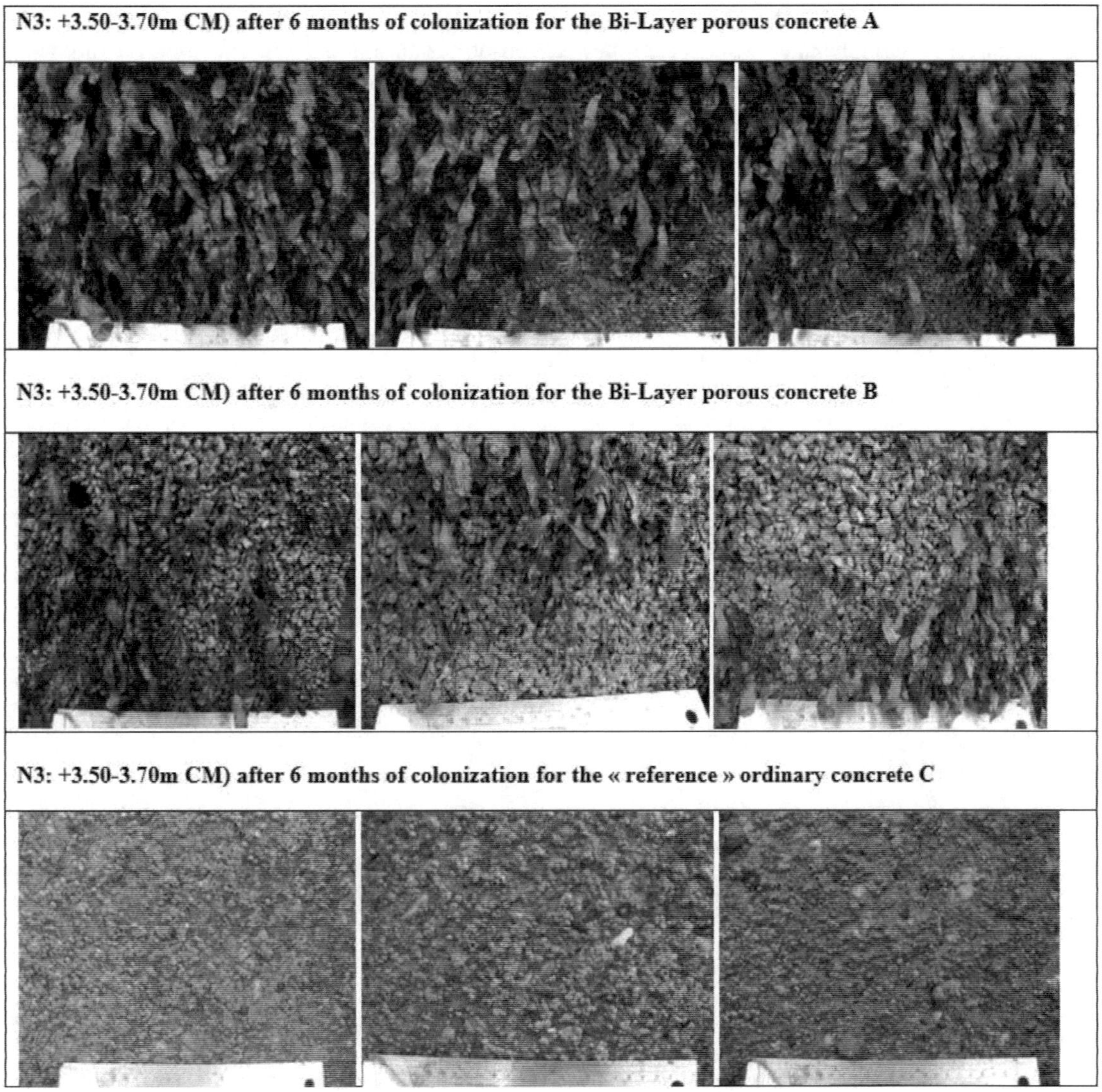

Figure 8. Pictures after 6 months showing the colonization for the different Bi-Layer porous concrete A, B and the “reference” ordinary concrete C

4. Conclusions

We developed a new laboratory test and method that is able to measure bio-receptivity of a concrete surface: this method is easy to implement in a laboratory and can be used favorably in a screening process to get the bio-receptivity of concrete surface in a relative short time (28 days). This work illustrates that it is necessary to get a surface that is highly porous where micro-organisms can find optimal place to settle and grow, but it remains difficult to define optimal concrete porous microstructure that will lead to very high bio-receptivity over time. We confirmed through the REXCOR and ANTICELTO projects that this laboratory test makes it possible to differentiate concrete formulations, in particular regarding the surface porosity parameter. This accelerated laboratory test could be a first step towards a standardization test for Bio-active concrete technologies.

References

[1] Bulleri F. and Chapman M.G., 'The introduction of coastal infrastructure as a driver of change in marine environments', *Journal of Applied Ecology* 47 (2010) 26-35.
[2] Perkol-Finkel S., Hadary T., Rella A., Shirazi R., Sella I., 'Seascape architecture – incorporating ecological considerations indesign of coastal and marine infrastructure', *Ecological Engineering* 120 (2018) 645–654
[3] Mitsch W.J and Jorgensen S.E., 'Ecological engineering: A field whose time has come', *Ecological Engineering* 20 (2003) 363-377.
[4] Mitsch W.J. 'What is ecological engineering', *Ecological engineering* 45 (2012) 5-12.
[5] Borsje B.W., Van Wesenbeeck, B.K. Dekker F., Paalvast P., Bouma T.J., Van Katwijk M.M.,De Vries M.B., 'How ecological engineering can serve in coastal protection', *Ecological Engineering* 37 (2011) 113-122.
[6] Pioch S., Kilfoyle F., Levrel H., Spieler R., 'Green marine construction' 61 (2011) 257-268.
[7] Pioch S., Saussola Ph., Kilfoyle K. and Spieler R., 'Ecological design of marine construction for socio-economic benefits: ecosystem integration of a pipeline in coral reef area', *Procedia Environmental sciences* 9 (2011) 158-152.
[8] Pioch S., Souche J.C., Le Saoût G., Clerc L., Bonato O., 'Comparaison expérimentale de la colonisation de béton bio-actifs en milieu marin', *Congress on artificial reefs : form materials to ecosystems*, ESITC Caen, (2014).
[9] Cuadrado-Rica H., 'Etude du comportement de bétons de coproduits coquilliers pour une utilisation en récifs artificiels', Thèse de l'université de Caen-Normandie (2016).
[10] Cresson P., Ruitton S., Harmelin-Vivien M., 'Artificial reefs do increase secondary biomass production: mechanisms evidenced by stable isotrope', *Marine ecology Progress Series* 509 (2014) 15-26.
[11] Baine M., 'Artificial reef: a review of the design, application, management and performance', *Ocean & Coastal management* 44 (2001) 241-259.
[12] Hayek M., Salgues M.,Souche J.C., Cunge E., Giraudel C. and Paireau O.,'Influence of the Intrinsic Characteristics of Cementitious Materials on Biofouling in the Marine Environment, *Sustainability* 13 (2021), 2625
[13] Perkol-Finkel S., Sella I. 'Ecologically Active Concrete for Coastal and Marine Infrastructure: Innovative Matrices and Designs' From Sea to Shore – *Meeting the Challenges of the Sea*, 2015
[14] Hayek M., Salgues M., Habouzit F., Bayle S., Souche J.C., De Weerdt K., Pioch S., 'In vitro and in situ tests to evaluate the bacterial colonization of cementitious materials in the marine environment', *Cement and Concrete Composites*, 113 (2020) 103748
[15] Dubois-Brugger I., Duchand S., Horgnies M., Morin V., DeLarrard F., Lecomte G., 'Concrete structure for marine and river application', *WO* 2018042240-A1 (2018).
[16] Sato K., Kozuki Y., Nishimura H., Tara C., Nakanishi T., Nakata H. Yamaguchi N., Nogami F., Yamanaka R., Tabohash T and Kobayashi M. 'Characteristics of Periphytic algae on concrete contained amino acid', 9th *International meeting on the Environmental management of enclosed coastal Seas (EMECS9)*, Baltimore, (2011)

Diversity, structures assemblages and production of benthic communities on artificial reefs, a comparative case study in the English Channel

B Vivier[1, 2], J C Dauvin[3], M Navon[1, 2], L Chasselin[1], M Deloor[1], A M Rusig[1, 2], I Mussio[1, 2], M Boutouil[4], J Salaün[3,5,6] and P Claquin[1, 2]

[1] Normandie Université, Université de Caen Normandie, F-14032, Caen, France
[2] Laboratoire Biologie des ORganismes et Ecosystèmes Aquatiques (BOREA, UMR CNRS 8067), Muséum National d'Histoire Naturelle, Sorbonne Université, Université de Caen Normandie, IRD 207, Université des Antilles. Centre de Recherches en Environnement Côtier (CREC) - Station Marine, BP49, 54, rue du Docteur Charcot - 14530 Luc-sur-Mer, France
[3] Laboratoire Morphodynamique Continentale et Côtière, UMR CNRS 6143 M2C, Normandie Université, Université de Caen Normandie, UNIROUEN, 24 rue des Tilleuls, F-14000 Caen, France
[4] Ecole Supérieure d'Ingénieurs des Travaux de la Construction de Caen (ESITC Caen), 14610 Epron, France
[5] Atlantique Landes Récifs, 1 rue du Pitey, ZA des deux pins, 40130 Capbreton
[6] Université Paul Valéry, Montpellier 3, Laboratoire d'Aménagement et de Géographie de Montpellier EA LAGAM, F34000, Montpellier, France

Email: pascal.claquin@unicaen.fr

Abstract. In this study, the diversity and structures assemblages of benthic communities present on artificial reefs (AR) immersed for 5 years were monitoring during a full year in 2020. The comparison of two different sites Bernières in the Bay of Seine and Cherbourg in the central part of the English Channel brings innovative results on the efficiency of such structures. Benthic fauna and macroalgae communities were studied; several biotic indices like Shannon-Wiener diversity index were calculated. Benthic fauna was classified according to their trophic group and the biomass was estimated. Our results pointed out strong differences for several indicators between sites and seasons. Benthic fauna was more abundant in the Bay of Seine and more diversified than in the Bay of Cherbourg. Primary producers' diversity and biomass were higher in the Bay of Cherbourg and dominated by Rhodophyceae species. Primary production results showed that the Bay of Cherbourg was a more productive system than the Bay of Seine. This study highlighted the efficiency of such structures to create habitats and promote biomass and diversity of associated living communities. In comparable conditions, different systems were highlighted: a "primary producer reef" and a "primary consumer reef".

1. Introduction

Artificial reefs (ARs) are man-made structures emplaced in aquatic environments that serve as habitats or shelters for organisms. AR have long been used to attract fish and the development of these structures has been intensified over the three last decades [1–3]. These structures are usually expected to produce an overall increase in species richness by protecting some species, and also an increase and diversification of trophic contributions [4–6]. The improvement in habitat is generally reflected in greater food availability and more shelter against predators, as well as new recruitment areas for juveniles of various species (benthic invertebrates or fish), which explains the increase in organism biomass associated with these structures after their installation [7,8]. The ecological value of artificial structures as habitats for native species can vary in relation to many structural and environmental factors [9–11]. In this study, the biodiversity, the trophic assemblages and the production associated to AR have been investigated. The comparison of identic structures in two different sites along the French coasts of the English Channel brings important indications on the elevated success rate of such projects in order to restore impacted ecosystems. This study also provides important understanding on the biodiversity – production relation. Moreover, the seasonal monitoring of several biological and environmental parameters over a full year after five years of immersion brings reliable data on the benthic communities (producers and consumers) associated to a stable hard substrate ecosystem.

2. Materials & Methods

2.1. Sampling sites

AR were deployed in two different macrotidal sites (Bay of Cherbourg, CHER and Bay of Seine in Bernières-sur-mer, BERN) in 2015. These AR were designed and immersed within the European project RECIF and were used as an example of artificial subtract in the European Interreg Va project MARINEFF (2018-2022). The first site CHER is characterised by a sandy bottom area while the second site BERN is characterised by sandy-rocky bottom area. In both sites, the low tide bathymetry is about 4 m. At CHER, the AR was made up of 12 modules grouped in 3 clusters to create a triangle structure and only 3 modules in BERN. Each module is 3 m long, 2 m wide and 1.35 m high (8,1 m^3 each module - 32,4 m^3 by cluster - 97,2 m^3 in total at CHER). The modules were constructed in three levels (Figure 1), the first level was a supporting base to avoid sinking; the second level was made up of solid concrete pillars and empty concrete drains; the third level was a metal cage filled with cinder blocks. The dimensions of each cinder blocks were 0.4 m long by 0.2 m wide by 0.2 m high. These cinder blocks were made with porous concrete.

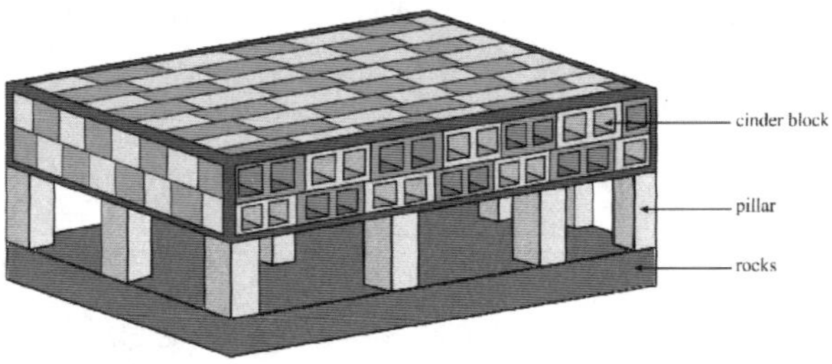

Figure 1. Representation of one AR module.

2.2. Sample collection

2.2.1. Benthic communities and production. The sampling of benthic communities was conducted from December 2019 to September 2020. Three campaigns were performed in BERN and four at CHER. At each campaign, six cinder blocks of the upper face of the modules were sampled randomly and taking back to the laboratory for analysis. In order to prevent any loss of fauna, each cinder block was carefully disposed in a planktonic net (500 µm) then closed and bring back to the boat. Each cinder block was conserved separately in a tank until the transport to the marine station of Luc-sur-mer. Then they were conserved in their planktonic net in several large basin with an open sea water renewal. Each cinder block was first incubated in a hermetic box with controlled lighting in the large basin in order to record the cinder block production. During each incubation, photoactive radiation (PAR) was recording with a RBR solo3 PAR logger connected to a Li-COR "Underwater Quantum Sensor" LI-192. Dissolved oxygen (mg O2.L-1) was recorded with an EXO Optical Dissolved Oxygen Sensor hermetically connected on the system. Each incubation consisted into seven different light conditions (dark 1 – PAR 1 – PAR 2 – PAR 3 – PAR 4 – PAR 5 – dark 2). Oxygen production was calculated as the slope of the linear regression of O2 concentration against incubated time per light condition and expressed in µmol O2.m-2.h-1, after integration of water volume contained in the box (58.8 L) and the cinder block surface. The gross primary production (GPP) was calculated for each PAR condition as the addition of the respiration (dark 2 condition) and the net primary production (NPP) of the selected condition. After the incubation, all macroalgae were sampled and conserved at -20°C for species identification and biomass estimation. The biomass of each species was determined and expressed Ash-Free Dry Weight (AFDW) per m^2 (loss of weight of dry organisms after 5h at 500°C). Ash Free Dry Weight biomass was converted to carbon content using a conversion factor of 0.35 [12,13]. Macroalgae species were assigned to their class (Ulvophyceae, Rhodophyceae or Phaeophyceae). Then, the cinder block was fixed in 10% formaldehyde, rinsed and scraped carefully in order to sample all organism which were preserved in 70% ethanol, identified to the lowest possible taxonomic level and counted (sieved on 0.5 mm diameter). The biomass of each taxonomic group was determined and expressed in terms of AFDW per m^2 (loss of weight of dry organisms after 5h at 500°C). Ash Free Dry Weight biomass was converted to carbon content using a conversion factor of 0.518 [14]. Benthic invertebrates were assigned to six trophic groups: grazers (Gr), filter feeders (FF), predators (Pr), scavengers – omnivorous (Scv/O), sub-surface deposit feeder (ssDF) and surface deposit feeder (sDF). The community structure analysis was based on the taxonomic richness, the Shannon-Wiener diversity index H′ and the Piélou evenness index J′. The taxonomic richness corresponded to the total number of taxa recorded on the six cinder blocks sampled at each campaign. Both indices, H' and J' using log2 were calculated for each campaign too.

2.2.2. Abiotic parameters and inorganic nutrient analyses. In the same time, high frequency measurements using the SMILE instrumented buoy located in the Bay of Seine (0°19'41.00"O 49°21'14.00"N) near BERN was collected (data available doi.org/10.17882/53689). Conductivity and temperature (tetracon sensor, WTW™), turbidity (Seapoint turbidity meter, Seapoint Sensor), oxygen (AADI Oxygen optode, Anderraa) and fluorescence (Cyclops-6K, Turner Design) were combined in an automatic NKE instrumentation (MP7, NKE Instrumentation®). PAR (photosynthetically active radiation) was measured with Saltantics sensor. For the inorganic nutrient analyses, a water sample was collected at each campaign in CHER and in autumn at BERN. Each sample was filtered using cellulose acetate filter (ClearLine, CA, 33mm, 0.45 µm) in 50 mL falcon tube and frozen (-20°C). Analysis were conducted using a Seal Analytical AA-3 system [15]. The limits of quantifications were 0.02 $\mu mol.L^{-1}$ for PO_4^{3-} and 0.05 $\mu mol.L^{-1}$ for NO_3^-, NO_2^- and $Si(OH)_4$. NH_4^+ were analysed by fluorometric measurements according to the protocol of Oriol et al. (2014). N_{tot} was calculated as follow: $N_{tot} = NO_3^- + NO_2^- + NH_4^+$.

2.3. Data treatment and analysis
Data analysis was performed with R i386 3.5.1 [17]. The non-parametric Kruskal-Wallis test was used to find differences in benthic macrofauna between sites and seasons in terms of species richness, abundance, biomass, Piélou's evenness index J' and Shannon-Wiener diversity index H'.

3. Results
3.1. Biotic indices and structure assemblages
A total of 40 macrofauna taxa was recorded, 35 in BERN and 37 in CHER, considering all the campaigns. Results showed that Shannon-Wiener diversity index and Piélou evenness index were significantly higher in CHER than at BERN at every season. These results indicate that CHER site hosted more diversified communities than BERN, (H') even if CHER seemed more dominated by one species in comparison with BERN (J'). Contrarily, the total biomass was significantly higher in BERN than CHER for all surveys. No significant differences were measured for taxonomic richness and abundance between the two sites and the different campaigns (Tables 1 and 2). In BERN during the winter, the most abundant taxa (> 30 individuals per m^2) were Amphipoda (86 per m^2), Aphroditidae (30 per m^2), *Pilumnus hirtellus* (178 per m^2) and *Pisidia longicornis* (90 per m^2). Spring was dominated by Amphipoda (230 per m^2), Aphroditidae (268 per m^2), Caprellidae (97 per m^2), Isopoda (70 per m^2), Nemertea (46 per m^2), Ophiuridae (190 per m^2), *Pilumnus hirtellus* (148 per m^2), *Pisidia longicornis* (84 per m^2), *Porcellana platycheles* (52 per m^2), Sabellidae (79 per m^2), Serpulidae (77 per m^2), Syllidae (44 per m^2) and Tanaidacea (91 per m^2). Autumn season was dominated by Amphipoda (239 per m^2),

Aphroditidae (102 per m^2), Bivalvia (140 per m^2), Caprellidae (35 per m^2), Isopoda (373 per m^2), Lysianassidae (33 per m^2), Ophiuridae (129 per m^2), *Pilumnus hirtellus* (270 per m^2), *Pisidia longicornis* (340 per m^2), *Porcellana platycheles* (40 per m^2), Sabellidae (83 per m^2), Serpulidae (58 per m^2), Tanaidacea (88 per m^2). In CHER, winter was dominated by Bivalvia (30 per m^2), Isopoda (47 per m^2), Ophiuridae (53 per m^2), summer by Amphipoda (42 per m^2), Cirratulidae (24 per m^2), Dynamenae (55 per m^2), Isopoda (38 per m^2), Nereidae (43 per m^2), *Pisidia longicornis* (70 per m^2). Autumn was dominated by Amphipoda (184 per m^2), Aphroditidae (40 per m^2), Ascidiacea (35 per m^2), Balanomorpha (112 per m^2), Bivalvia (121 per m^2), Capitellidae (69 per m^2), Cirratulidae (67 per m^2), Dynamenae (48 per m^2), Ectoprocta (96 per m^2), Eunicidae (219 per m^2), Gastropoda (39 per m^2), Isopoda (124 per m^2), Lysianassidae (49 per m^2), Ophiuridae (56 pe m^2), Phyllodocidae (30 per m^2), *Pilumnus hirtellus* (59 per m^2), *Pisidia longicornis* (48 per m^2), Porifera (35 per m^2), Sabellidae (32 per m^2) and Serpulidae (35 per m^2). Finally, winter was dominated by Ascidiacea (50 per m^2), Eunicidae (43 per m^2), Ophiura (64 per m^2) and Sabellaria (31 per m^2).

Table 1. Main characteristics of macrofauna with different univariate indices for both sites and all seasons. H': Shannon-Wiener diversity index; TR: Taxonomic Richness; J': Piélou's evenness index; A: Abundance (number of individuals.m^{-2} and B: biomass (g AFDW.m^{-2}). Significant differences (p-value < 0.05) are indicated by letter (a, b).

		Cherbourg (CHER)				Bay of Seine (BERN)		
		Winter	Spring	Summer	Autumn	Winter	Spring	Autumn
Macrofauna	H'	3.95^a	4.17^a	4.24^a	4.45^a	3.10^b	3.77^b	3.69^b
	TR	26	27	31	33	28	30	29
	J'	0.84^a	0.88^a	0.86^a	0.88^a	0.65^b	0.77^b	0.76^b
	A	400	374	490	1071	524	1563	2078
	B	1.49^a	1.06^a	1.62^a	2.39^a	3.75^b	4.6^b	2.5^b

Table 2. Results of the non-parametric Kruskal-Wallis test on five parameters between the two sites. H': Shannon-Wiener diversity index; TR: Taxonomic Richness; J': Piélou's evenness index; A: Abundance; B: biomass (g AFDW.m^{-2}); Df: Degrees of Freedom; F: F-ratio score; BERN: Bay of Seine; CHER: Bay of Cherbourg.

	Df	p-value	Interpretation
H'	1	0.03389	BERN≠CHER
TR	1	1	BERN=CHER
J'	1	0.03389	BERN≠CHER
A	1	0.0771	BERN=CHER
B	1	0.03389	BERN≠CHER

The 40 taxa were assigned to six trophic groups. Abundance and biomass of these trophic groups on both sites and during the entire survey are represented in the figure 1. In BERN, the total abundance was higher in autumn than in the rest of the year (Figure 2A). In winter, filter feeders and predators were the most abundant trophic groups (143 and 193 individuals per m^2). In spring, the most abundant group was scavengers – omnivorous benthic invertebrates (590 individuals per m^2) followed by filter feeders one (296 individuals per m^2). Finally, in autumn the most abundant group was the filter feeders (523 individuals per m^2) immediately followed by the surface deposit feeders (507 individuals per m^2). As for BERN, the total abundance of benthic invertebrates was higher in autumn for CHER (Figure 2B). The variations in term of abundance from winter to summer were non-significant in CHER. However, the abundance of filter feeders, predators and sub-surface deposit feeders reached respectively 358, 397 and 564 individuals per m^2 in autumn. Some significant differences could be noted between the two sites, especially for spring were the abundances of some trophic groups like scavengers – omnivorous or filter feeders were three to six times higher in BERN. In term of biomass in BERN (Figure 2C), filter feeders were the main contributors with 56% of the total AFDW biomass in winter, 84% in spring and 76% in autumn. Predators represent 38% of the total AFDW biomass in winter, 10% in spring and 5% in autumn. Some trophic groups like scavengers – omnivorous or surface deposit feeders represent a very small part of the total AFDW biomass in comparison with their large abundances. In CHER, the main contributors were filter feeders with 27% of the total AFDW biomass in winter, 48% in spring, 56% in summer and 29% in autumn. Predators represent also a large part of the total biomass with 63% in winter, 41% in spring, 38% in summer and 17% in autumn. Moreover, sub-surface deposit feeders represented the higher percentage of total biomass in Autumn (39%) against 6% in the Bay of Seine at the same period.

A total of 18 different macroalgae species was recorded, 9 in BERN and 15 in CHER considering all campaigns. *Cryptopleura ramosa, Dictyota dichotoma, Heterosiphonia plumosa, Hypoglossum hypoglossoides, Kallymenia reniformis, Plocamium cartilagineum, Sphaerococcus coronopifolius, Sphondylothamnion multifidum* and *Ulva* sp. were recorded in The Bay of Seine. *Ahnfeltiopsis devoniensis, Bornetia secundiflora, Calliblepharis jubata, Cladophora pellucida, Cryptopleura ramosa, Halurus flosculosus, Hypoglossum hypoglossoides, Kallymenia reniformis, Laminaria digitata, Plocamium cartilagineum, Sargassum muticum, Sphaerococcus coronopifolius, Sphondylothamnion multifidum, Vertebrata byssoides* and *Vertebrata fruticulosa* were recorded in CHER. The table 3 summarise the AFDW biomass ($gC.m^{-2}$) for each macroalgae class in CHER and BERN. The total biomass was higher in CHER than in BERN at each season. Macroalgae community in CHER was dominated by Rhodophyceae (between 1.253 and 1.464 $gC.m^{-2}$) and some Ulvophyceae and Phaeophyceae were recorded in autumn (respectively 0.292 and 0.162 $gC.m^{-2}$). In CHER, macroalgae community was also dominated by Rhodophyceae but in higher quantities (between 14.617 $gC.m^{-2}$ in winter and 64.743 $gC.m^{-2}$ in summer). Moreover, Phaeophyceae were recorded at each campaign and some Ulvophyceae in early winter (0.011 $gC.m^{-2}$).

3.2. Primary production

The figure 3 shows the incubations results. The GPP and NPP were calculated at saturation light and the respiration was obtained from dark condition at the beginning of each incubation. The saturation light obtained for the BERN AR incubation was 933 µmol photons m^{-2} s^{-1} and 714 µmol photons m^{-2} s^{-1} for CHER. NPP incubation results were strongly different between both sites. In BERN, between spring and autumn, the total respiration was negative with -2,788 +/- 972 µmol $O_2.m^{-2}.h^{-1}$ and -1,706 +/- 1,168 µmol $O_2.m^{-2}.h^{-1}$ respectively. The consequence of this strong oxygen consumption in BERN was a negative GPP for all seasons. Indeed, the GPP decreased between winter and autumn from 3,512+/- 7,205 µmol $O_2.m^{-2}.h^{-1}$ to 1,165 +/- 1,659 µmol $O_2.m^{-2}.h^{-1}$. There was no production measured during spring season for any of the six incubated AR. In CHER, the NPP was always positive. It showed a decrease between winter and spring from 12,907 +/- 5,958µmol $O_2.m^{-2}.h^{-1}$ to 695 +/- 2,965 µmol $O_2.m^{-2}.h^{-1}$ before reached 16,344 +/- 7,632 µmol $O_2.m^{-2}.h^{-1}$ in summer and 13,875 +/- 4,555 µmol $O_2.m^{-2}.h^{-1}$ in Autumn. The total respiration was almost null between winter and spring (-1,484 +/- 371 µmol $O_2.m^{-2}.h^{-1}$ and 283 +/- 93 µmol $O_2.m^{-2}.h^{-1}$ respectively) and reached -3,174 +/- 769 µmol $O_2.m^{-2}.h^{-1}$ and -2,336 +/- 615 µmol $O_2.m^{-2}.h^{-1}$ in summer and autumn. Contrarily to BERN, the GPP in CHER was always positive. Between winter and spring, it decreased from 14,391 +/- 6,256 µmol $O_2.m^{-2}.h^{-1}$ to 978 +/- 2,918 µmol $O_2.m^{-2}.h^{-1}$. It increased until 19,518 +/- 7,748 µmol $O_2.m^{-2}.h^{-1}$ in summer and stay stable at 16,212 +/- 4,735 µmol $O_2.m^{-2}.h^{-1}$ in autumn.

Table 3. AFDW biomass in g C.m^{-2} for each identified macroalgae class (Ulvophyceae, Phaeophyceae and Rhodophyceae) in the Bay of Seine (BERN) and the Bay of Cherbourg (CHER) for each campaign.

		AFDW biomass (gC.m^{-2})		
Site	Season	Ulvophyceae	Phaeophyceae	Rhodophyceae
BERN	Winter	0	0	1.464
	Spring	0	0	1.418
	Autumn	0.292	0.162	1.253
CHER	Winter (February)	0	1.069	14.617
	Spring	0	0.283	25.360
	Summer	0	4.223	64.743
	Autumn	0	0.493	36.575
	Winter (November)	0.011	0.478	41.130

IOP Conf. Series: Materials Science and Engineering 1245 (2022) 012003 doi:10.1088/1757-899X/1245/1/012003

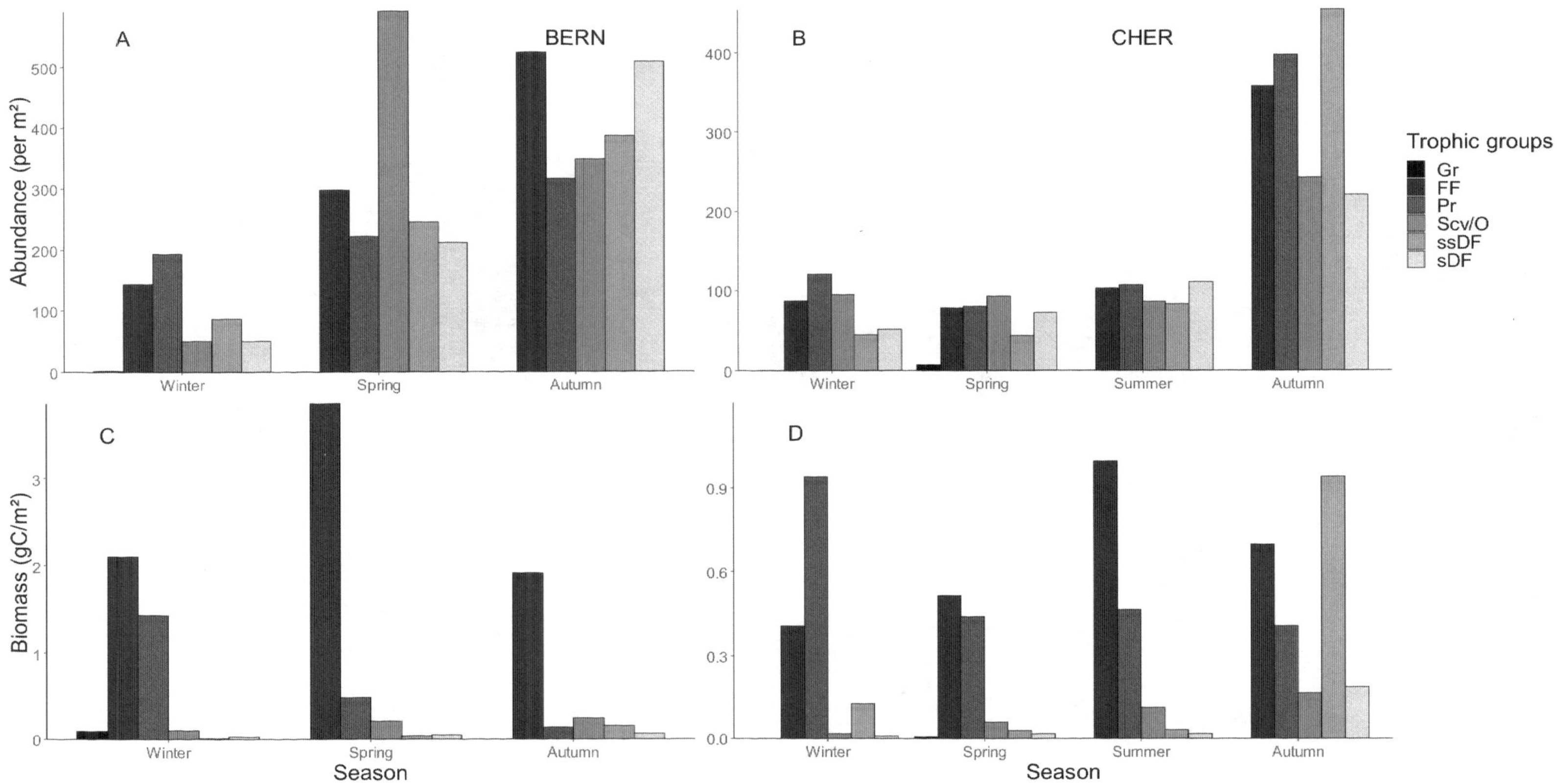

Figure 2. Abundance (A, B) and percentage of total AFDW biomass per season (C, D) of trophic groups in the Bay of Seine (A, C) and the Bay of Cherbourg (B, D) for every season. Trophic groups: grazers (Gr), filter feeders (FF), predators (Pr), scavengers – omnivorous (Scv/O), sub-surface deposit feeder (ssDF) and surface deposit feeder (sDF).

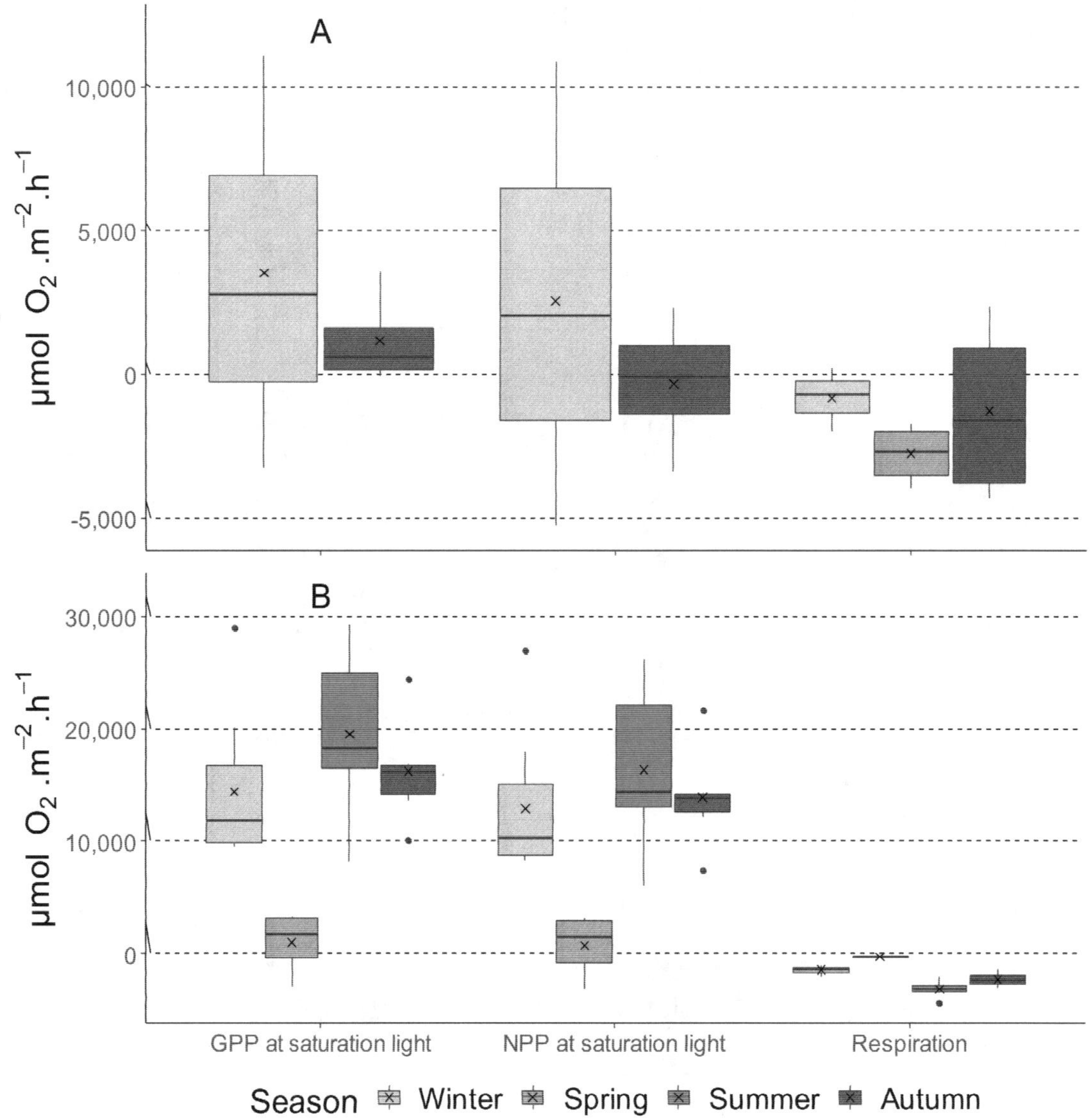

Figure 3. Gross primary production (GPP), respiration and net primary production (NPP) incubation results in μmol $O_2.m^{-2}.h^{-1}$ per cinder block at saturation light and at each season in BERN (A) and CHER (B).

The table 4 showed the monitored environmental parameters during the survey.

Table 4. Environmental parameters at each site during the survey.

	BERN			CHER			
	Winter	Spring	Autumn	Winter	Spring	Summer	Autumn
Temperature (°C)	10	14	20		14	16	17
Salinity	33.36	33.26	33.39	33.40	34.30	34.60	34.56
PAR (µmol photons m^{-2} s^{-1})	8	134	43				
pH	7.97	7.99	7.88				
Turbidity (NTU)		5.00	88.40			0.70	0.60
Dissolved oxygen (mg/L)	6.23	6.45	5.24				
MES (mg/L)	13.80	4.43	1.57				
Chlorophyll *a* (µg/L)	0.66	4.36	1.05		1.33	1.54	0.87
NO_{TOT} (µmol/L)	24.88	9.00	2.16	11.81	2.20	0.17	3.31
PO_4^{3-} (µmol/L)	0.81	0.75	0.12	0.53	0.03	0.03	0.30
$SiOH_4$ (µmol/L)	15.64	1.09	6.11	5.30	1.52	1.41	3.39
$NH4^+$ (µmol/L)	1.36	0.94	0.80				

4. Discussion

Firstly, experimental sites showed high environmental differences such as hydrodynamism, tide amplitude, currents velocity or water turbidity. Indeed, CHER is a much more protected area than BERN due to the long dike. One of the consequences of this manmade structure is that the hydrodynamism and the tide influence on local currents is completely different and reduced in comparison with the open bay of Seine. Secondly, the water parameters were also very different between both sites. BERN is located in the bay of Seine under rivers influence where abiotic parameters showed important variations. The water turbidity is much higher in BERN than in CHER, inorganic nutrients concentration is also strongly higher in BERN than in CHER. Considering these disparities, the comparison of several biotic variables on identic AR immersed in both sites since 2015, it is interesting to study the benthic communities' assemblages and successions. Several structural factors such as depth, orientation, age and surface complexity may influence the structure and the diversity of benthic communities [1,11,18,19]. In order to eliminate some of these factors like orientation, the sampling was carried out on each artificial reef cluster in both sites.

The benthic fauna recorded in CHER showed a more diverse community than in BERN (H'). The total biomass was significantly higher in BERN than in the CHER. In order to have a functional approach, benthic fauna was assigned to several trophic groups. For both sites, the maximal of abundance was reached in autumn as illustrated by the figure 2. In term of biomass, they were high differences between sites and seasons. Indeed, in BERN, the total biomass per season was always dominated by filters feeders and, in smaller proportions, predators in winter. In CHER, the benthic fauna was also dominated by filters feeders and predators in spring and summer. The benthic community was dominated by sub-surface deposit feeders in autumn (followed by filters feeders and predators) and by predators in winter. These results indicated that the trophic webs in both sites were very different with a more complex system with distinct seasonal successions in CHER. Moreover, at the same season, the dominant trophic groups were different between both sites. For example, in autumn, the benthic fauna community in BERN was completely dominated by filters feeders and in CHER, a more diversified structure was assessed with sub-surface deposit feeders, filters feeders and predators. These ecosystems were considered as stable because of the long immersion time of the AR (from 2015). The only two different ways to avoid over-competition between marine organisms is the food source separation and habitat specialisation. Considering that the immersed structures were the same, the recorded differences of fauna trophic structure between sites may be induced by abiotic factors (hydrodynamism, salinity, temperature…) and surrounding communities present before the artificial reef immersion. Water temperature is known to influence the metabolism of a large part of marine organisms and may explained the lower abundance for every trophic groups recorded on both sites in winter [20–22]. These results tend to confirm the efficiency of such structures to create novel habitats and restore hard substrate communities. Indeed, our results showed high similarities with natural reef investigations in this region indicating that AR immersion reproduced natural reef ecosystem [23].

The investigation of relations between benthic fauna and macroalgae in AR systems have been poorly investigated [24]. The results obtained on macroalgae communities showed a significant higher diversity in CHER than in BERN (15 against 8 species in total). In BERN, the diversity and total biomass were higher in autumn in comparison with the rest of the year. Ulvophyceae and Phaeophyceae species were recorded only in autumn on this site. In CHER, total biomass and diversity were higher in summer, Ulvophyceae species were recorded only in winter. Moreover, *Sargassum muticum*, an invasive species, was recorded almost all the year in CHER and never in BERN. These results indicated that CHER AR supports a more diversified primary producers' community as for the benthic fauna. The structural differences between macroalgae communities on both sites may be explained by abiotic factors like turbidity, available PAR or inorganic nutrient availability. As for benthic fauna communities, macroalgae may be influenced by the structure, material and surface heterogeneity of the AR which cannot be the case here because of the AR homogeneity between sites [25,26]. However, even if the inorganic nutrient concentration especially N_{tot} were higher BERN than in CHER, the higher biomass and diversity were recorded in CHER. These results may be explained in part by the elevated water turbidity in the BERN and by the strong tidal currents in comparison with CHER [27,28]. Macroalgae communities were strongly dominated by *Cryptopleura ramosa* for BERN AR and by *Kallymenia reniformis* and *Sphaerococcus coronopifolius* for CHER. Several species recorded during the monitoring, such as *Calliblepharis jubata, Heterosiphonia plumosa* or *Sacchoriza polyschides* in CHER correspond to protected area indicative species of Atlantic reef ecosystems. However, the dominant ones correspond to exposed area indicative species of Atlantic reef ecosystem (i. e. *Kallymenia reniformis*) [23]. The large majority of these species corresponds to hard substrates habitat species; however, CHER is an area dominated by sandy seabed and a strong proportion of the macroalgae recorded in this study were not considered before as determinant species for this area. These original results for this area needed to be considered for future environmental assessment in CHER and dominant species recorded in this study may be considered as determinant ones. Generally, these results tend to confirm the good efficiency of these structure to restore or create novel habitats because recorded macroalgae communities corresponds to those observed in near rocky areas [29]. Non-indigenous species establishment in AR associated ecosystem have been largely studied and the immersion of such structures may promote their colonization [30,31]. *Sargassum muticum* is originated from Asia and already well implanted in the English Channel [32,33]. Its presence all the year at CHER and its absence in BERN was surprising and difficulted to explain because of its abundance along the English Channel coast.

The primary production results obtained by incubations (at saturation light) showed significant differences between sites and seasons. Not surprisingly because of the higher macroalgae biomass, AR in CHER were the most productive ones. In CHER, the higher production was recorded during the summer. This result corresponds to the season with higher macroalgae biomass and diversity on this site. The GPP reached 19,518 +/- 7,748 $\mu mol\ O_2.m^{-2}.h^{-1}$ in summer in CHER, in comparison the maximal GPP in BERN was reached in autumn and was negative. The maximal respiration was recorded in summer in CHER and also correspond to the most productive season on this site. The autumn in CHER showed also a high GPP and high oxygen consumption. However, a relatively low macroalgae diversity was recorded in autumn on this site (3 different species). The higher respiration results recorded in summer and autumn for CHER corresponded to the seasons with the higher diversity index (H') and total benthic fauna biomass. These results showed that the associated AR ecosystem in CHER was more productive than BERN for each season. The relatively poor macroalgae biomass in the BERN was not sufficient to compensate the consumed oxygen, especially during the winter season were the NPP and respiration were strongly negative. The lower saturation light for CHER incubations (714 µmol photons m^{-2} s^{-1}) in comparison with BERN (933 µmol photons m^{-2} s^{-1}) indicated that CHER communities were more able to realise the photosynthesis at lower PAR. These results may be induced by the important Rhodophyceae community in CHER, indeed, these macroalgae are known for high photosynthetic efficiency at low light levels and may contribute to the high oxygen production measured for this site [34,35]. These results suggested a better photoacclimatation of CHER primary producers' communities and the needed of higher PAR quantity for BERN. In the same sea, at the same depth, in the same area, but in different water masses, two types of AR functioning were highlighted: One is a "primary producer reef" rich in macroalgae, CHER, and the second is a "primary consumer reef" where production is based on phytoplankton resources, BERN.

5. Conclusion

This study provided innovative results on the biodiversity and the trophic structure associated to AR immersed since several year in two different ecosystems of the English Channel. Our results showed important and diverse benthic fauna communities for both AR sites. The Bay of Cherbourg showed higher diversity and biomass of primary producers, dominated by Rhodophyceae macroalgae. However, the BERN AR showed a more diversified and abundant benthic fauna community than CHER. Primary production results showed significant differences between site; indeed, AR in CHER were more productive than BERN possibly because of the Rhodophyceae abundance on this site. In comparable conditions, two different ecological functioning were highlighted for identical AR structures, indeed, a strong "primary producer AR" were described in CHER and a "primary consumer AR" BERN. More generally, our results showed the great efficiency of such structures to create or restore habitats and promote biomass and diversity of associated living communities.

References

[1] Baine M 2001 Artificial reefs: a review of their design, application, management and performance *Ocean & Coastal Management* **44** 241–59
[2] Santos L N, García-Berthou E, Agostinho A A and Latini J D 2011 Fish colonization of artificial reefs in a large Neotropical reservoir: material type and successional changes *Ecological Applications* **21** 251–62
[3] Lowry M B, Glasby T M, Boys C A, Folpp H, Suthers I and Gregson M 2014 Response of fish communities to the deployment of estuarine artificial reefs for fisheries enhancement *Fisheries Management and Ecology* **21** 42–56
[4] Bodilis P, Seytre C, Charbonnel E and Francour P 2011 Monitoring of the Artificial Reef Fish Assemblages of Golfe Juan Marine Protected Area (france, North-Western Mediterranean) *Braz. J. Oceanogr.* **59** 167–76
[5] Hixon M and Beets J 1989 Shelter Characteristics and Caribbean Fish Assemblages - Experiments with Artificial Reefs *Bull. Mar. Sci.* **44** 666–80
[6] Piazza B P, Banks P D and La Peyre M K 2005 The Potential for Created Oyster Shell Reefs as a Sustainable Shoreline Protection Strategy in Louisiana *Restoration Ecology* **13** 499–506
[7] Perkol-Finkel S, Hadary T, Rella A, Shirazi R and Sella I 2018 Seascape architecture – incorporating ecological considerations in design of coastal and marine infrastructure *Ecological Engineering* **120** 645–54
[8] Sherman R 2002 Artificial reef design: void space, complexity, and attractants *ICES Journal of Marine Science* **59** S196–200

[9] Bortone S A 2006 A Perspective of Artificial Reef Research: The Past, Present, and Future *Bulletin of Marine Science* **78** 9
[10] Perkol-Finkel S, Zilman G, Sella I, Miloh T and Benayahu Y 2008 Floating and fixed artificial habitats: Spatial and temporal patterns of benthic communities in a coral reef environment *Estuarine, Coastal and Shelf Science* **77** 491–500
[11] Vivier B, Dauvin J-C, Navon M, Rusig A-M, Mussio I, Orvain F, Boutouil M and Claquin P 2021 Marine artificial reefs, a meta-analysis of their design, objectives and effectiveness *Global Ecology and Conservation* e01538
[12] Pihl L, Magnusson G, Isaksson I and Wallentinus I 1996 Distribution and growth dynamics of ephemeral macroalgae in shallow bays on the Swedish west coast *Journal of Sea Research* **35** 169–80
[13] Surif M B and Raven J A 1990 Photosynthetic gas exchange under emersed conditions in eulittoral and normally submersed members of the Fucales and the Laminariales: interpretation in relation to C isotope ratio and N and water use efficiency *Oecologia* **82** 68–80
[14] Brey T 1999 A collection of empirical relations for use in ecological modelling *Naga the ICLARM quarterly* **22** 24–8
[15] Aminot A and Kérouel R 2007 *Dosage automatique des nutriments dans les eaux marines: méthodes en flux continu* (Editions Quae)
[16] Oriol L, Garcia N and Cariou T 2014 Dosage de l'azote ammoniacal en milieu marin par fluorimétrie, Procédure : Protocole National. *SOMLIT* 1–14
[17] R Development Core Team 2008 *R: A Language and Environment for Statistical Computing* (Vienna, Austria: R Foundation for Statistical Computing)
[18] Bellou N, Papathanassiou E, Dobretsov S, Lykousis V and Colijn F 2012 The effect of substratum type, orientation and depth on the development of bacterial deep-sea biofilm communities grown on artificial substrata deployed in the Eastern Mediterranean *Biofouling* **28** 199–213
[19] Jaxion-Harm J and Szedlmayer S T 2015 Depth and Artificial Reef Type Effects on Size and Distribution of Red Snapper in the Northern Gulf of Mexico *North American Journal of Fisheries Management* **35** 86–96
[20] Fraser K P P and Rogers A D 2007 Protein metabolism in marine animals: The underlying mechanism of growth *Advances in Marine Biology, Vol 52* vol 52, ed D W Sims (San Diego: Elsevier Academic Press Inc) pp 267–362
[21] Hall A E, Herbert R J H and Stafford R 2021 Temporal and spatial variation in adult and juvenile mobile fauna associated with natural and artificial coastal habitats *Mar Biol* **168** 19
[22] Huey R B and Kingsolver J G 2011 Variation in universal temperature dependence of biological rates *Proc. Natl. Acad. Sci. U. S. A.* **108** 10377–8
[23] Anon 2004 *Connaissance et gestion des habitats et des espèces d'intérêt communautaire, tome 2: habitats côtiers*
[24] Yang X, Lv H, Li W, Guo M and Zhang X 2018 Effect of water motion and microhabitat preferences on spatio-temporal variation of epiphytic communities: a case study in an artificial rocky reef system, Laoshan Bay, China *Environmental Science and Pollution Research* **25** 12896–908
[25] Christie H, Jørgensen N M and Norderhaug K M 2007 Bushy or smooth, high or low; importance of habitat architecture and vertical position for distribution of fauna on kelp *Journal of Sea Research* **58** 198–208
[26] Wahl M 2009 *Marine hard bottom communities: patterns, dynamics, diversity, and change* (Berlin: Springer)
[27] Baux N, Murat A, Faivre Q, Lesourd S, Poizot E, Mear Y, Brasselet S and Dauvin J-C 2019 Sediment dynamic equilibrium, a key for assessing a coastal anthropogenic disturbance using geochemical tracers: Application to the eastern part of the Bay of Seine *Cont. Shelf Res.* **175** 87–98

[28] Cailleaud K, Forget-Leray J, Peuhiet L, LeMenach K, Souissi S and Budzinski H 2009 Tidal influence on the distribution of hydrophobic organic contaminants in the Seine Estuary and biomarker responses on the copepod Eurytemora affinis *Environ. Pollut.* **157** 64–71

[29] Lemesle S 2015 *Bioindicateurs du milieu marin : utilisation des macroalgues comme outil d'évaluation de la qualité biologique des eaux marines : réponse à une problématique d'échouages d'algues* These de doctorat (Caen)

[30] Glasby T M, Connell S D, Holloway M G and Hewitt C L 2007 Nonindigenous biota on artificial structures: could habitat creation facilitate biological invasions? *Marine Biology* **151** 887–95

[31] Morris R L, Golding S, Dafforn K A and Coleman R A 2018 Can coir increase native biodiversity and reduce colonisation of non-indigenous species in eco-engineered rock pools? *Ecological Engineering* **120** 622–30

[32] Kim H-G, Hawkins L E, Godbold J A, Oh C-W, Rho H S and Hawkins S J 2019 Comparison of nematode assemblages associated with Sargassum muticum in its native range in South Korea and as an invasive species in the English Channel *Mar. Ecol.-Prog. Ser.* **611** 95–110

[33] Strong J A, Dring M J and Maggs C A 2006 Colonisation and modification of soft substratum habitats by the invasive macroalga Sargassum muticum *Mar. Ecol.-Prog. Ser.* **321** 87–97

[34] Eggert A and Wiencke C 2000 Adaptation and acclimation of growth and photosynthesis of five Antarctic red algae to low temperatures *Polar Biol* **23** 609–18

[35] Weykam G and Wiencke C 1996 Seasonal photosynthetic performance of the endemic antarctic red algaPalmaria decipiens (Reinsch) Ricker *Polar Biol* **16** 357–61

IOP Conf. Series: Materials Science and Engineering 1245 (2022) 012004 doi:10.1088/1757-899X/1245/1/012004

Ecosystem approach of artificial reef through trophic web modelling

A Raoux[1], J Salaun[1,2,3], J P Pezy[1], S Pioch[3], J C Dauvin[1]

[1] Normandie Univ., UNICAEN, CNRS UMR 6143 M2C, Laboratoire Morphodynamique Continentale et Côtière, UMR CNRS 6143 M2C, 24 rue des Tilleuls, F-14000 Caen, France
[2] Association RA Landes
[3] Bio-geography, CNRS - UPV, UMR 5175 CEFE, University Montpellier 3, route de Mende, 34199 Montpellier cedex 5, France

Abstract. Along the French coast of the Atlantic and English Channel in a high tidal regime, only a limited number of Artificial Reefs (ARs) had been immerged in very few localities in the aim to increase the local production of fisheries of such hard artificial substratum. Moreover, ARs design have been planned to increase the diversity, abundance and biomass of benthic species associated to hard-bottom. Does the deployment of such artificial structure improve the production and ecological functioning on such equipped areas? Trophic models describing the interaction between species at different trophic levels and based on the quantification of energy flow and matter in ecosystems are able to respond to this question. They allow the application of numerical methods for the characterization of emergent properties of the ecosystem, also called Ecological Network Analysis (ENA). Usually, these indices have been proposed as ecosystem health indicators as they have been demonstrated to be sensitive to different impacts on marine ecosystems. Here, Ecopath ecosystem models composed of 23 compartments, from phytoplankton to mammals, are built to describe the situation “Before” and “After” the implantation of ARs in two areas: the Rade of Cherbourg in the central part of the English Channel along the French coast, and in the south of the Bay of Biscay along the Landes coast. In addition, ENA indices are calculated for the two periods, “before” and “after” the implantation of artificial reefs to compare network functioning and the overall structural trophic web properties. In both areas, increase of benthic and fish compartments insure an increase of the system maturity; nevertheless, the high surge of planktinovorous fishes on the Landes AR showed a protection role of such AR than changes of planktonic preys.

1. Introduction

Artificial reefs (ARs) are human made underwater structure placed at the bottom of the sea to mimic the natural reef physical and biological functions [1]. ARs are considered as an essential tool for increasing fishery production, providing recreational fishing and diving sites and provide protection against illegal trawling [2]. For more than fifty years, ARs have been deployed in France to respond to the decline in fish stocks [3]. The main aim of these structures is to sustain artisanal fisheries and enhance fish stocks [4]. Assessments of their effectiveness, when they have been carried out, have been focused only on certain ecological components such as commercial fishes [5]. However, Salaun et al. [5] pointed out the lack of ARs monitoring. This lack of feedback raises questions regarding the real ecological role of these structures [5]. Quantitative trophic web models can be used for this purpose since they describe the interactions between species at different trophic levels as they are based on the quantification of flows of energy and matter in ecosystems. Among these different existing approaches, the Ecopath with Ecosim (EwE) model has been intensively used and developed over the last three decades [6] and was recognised by the US National Oceanographic and Atmospheric Administration

(NOAA) as one of the ten biggest scientific breakthroughs in its 200 years history [7]. This modelling approach was developed to evaluate ecosystem-based management of fisheries and allowed the application of numerical methods for the characterization of emergent properties of the ecosystem, also called Ecological Network Analysis (ENA) [8]. These joint analyses had been frequently applied to coastal and marine systems to assess changes in their functioning in response to environmental perturbations [9]. Here, Ecopath ecosystem models composed of 23 compartments, from phytoplankton to mammals, are built to describe the situation "Before" and "After" the implantation of ARs in two areas: the Rade of Cherbourg in the central part of the English Channel along the French coast, and in the south of the Bay of Biscay along the Landes coast. In addition, ENA indices are calculated for the two periods, "before" and "after" the implantation of artificial reefs to compare network functioning and the overall structural trophic web properties.

2. Materials and Methods

2.1 Study areas

Two artificial reef located along the French Atlantic coast were selected: the Rade of Cherbourg in the North Cotentin in the central part of the English Channel along the French coast, and the Capbreton concession in the south of the Bay of Biscay along the Landes coast (Figure 1).

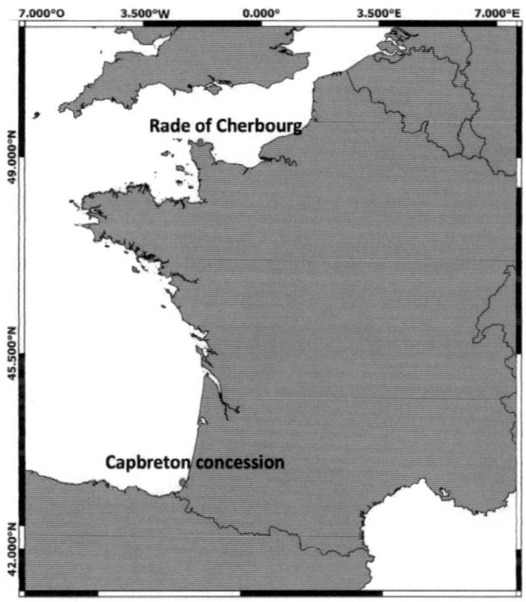

Figure 1. Locations of the Artificial Reefs of the Rade of Cherbourg and of the Capbreton concession.

Under the framework of the European project RECIF, an artificial reef system was immersed in the Rade of Cherbourg on April 2015. The experimental artificial reef was composed of 12 modules (Figure 2) grouped in 3 clusters to create a triangle structure. Each module was 3m long, 2m wide and 1.35m high (8,1m^3 each module - 32,4m^3 by cluster - 97,2m^3 in total). The RECIF project aimed to reuse marine by-products to create artificial reefs for a better management of marine resources. To do this, the RECIF project used waste shells from the shellfish industry to create artificial reef systems. Monitoring of colonisation of this experimental AR are currently being conducted under the framework the European MARINEFF project.

Figure 2. Immersion of a module on April 2015 in the Rade of Cherbourg (ESITC, Caen).

In the south of Bay of Biscay, ARs have been deployed 2km south off the head of the Canyon and 2.2km off the coast on sandy bottom at 20m depth and 20km away from natural reef [19](Figure 3). Three types of artificial reefs have been deployed by the association « Atlantique Landes Récifs »: grapes of concrete bonna pipes deployed in 1999 on three sites around 200m² each, the module “Typi” in 2010 with a 11m² footprint and the “Babel” in 2015 with a 5m² footprint (Figure 3).

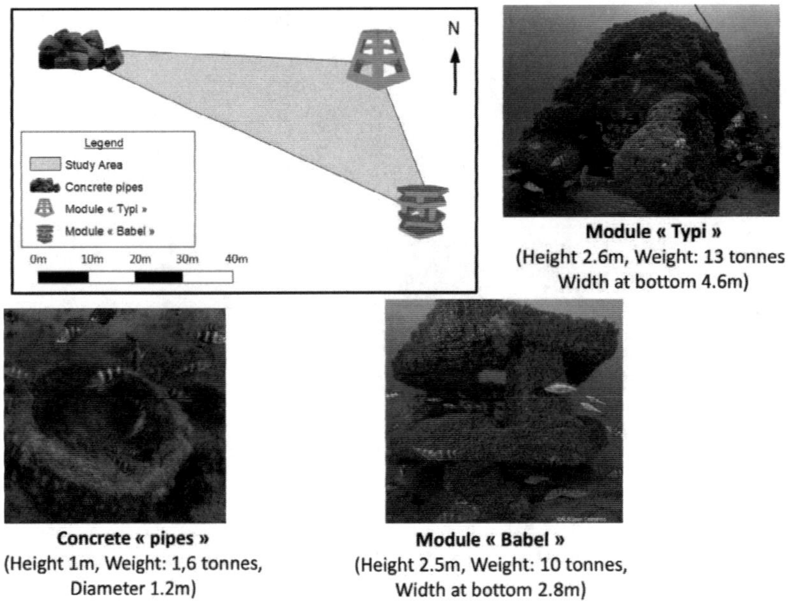

Figure 3. Capbreton concession with the three artificial reefs

2.2 Ecopath modelling approach

In order to gain further knowledge on the ecosystem structure and functioning before and after the implantation of ARs in the two areas: Rade of Cherbourg and in the Capbreton concession, trophic web models were built. The Ecopath with Ecosim (EwE) approach [20] was used to model the trophic webs at the two sites. Thus, three Ecopath models were built: two baseline Ecopath models called “Before Artificial Reef _Cherbourg” (BAR_Cherbourg) and “Before Artificial Reef_Landes”

(BAR_Landes) and one model called Artificial Reef_Landes (AR_Landes) based on the data collected on the three artificial reefs of the Capbreton concession.
Ecopath is a mass-balance, single-solution model that uses linear equations to estimate flows between numbers of functional groups established a priori [8]. The parameterization of an Ecopath model is based on satisfying two equations. The first equation (Eq. 1) describes the production for each compartment in the system as a function of the consumption to biomass ratio (Q/B) of its predators (j), the fishing mortality (Yi, gC·m−2), the net migration (Ei; emigration – immigration, year−1), the biomass accumulation (BAi, year−1) and its natural mortality (1—EEi). The Ecotrophic Efficiency (EE) is the fraction of total production that is consumed in the system (by fishing activity or by predators). Its value can never exceed 1. (1-EEi) represents the fraction of mortality not explained by the model, such as mortality due to old age or diseases.

$$B\left(\frac{P}{B}\right)_i = \sum_j B_j \left(\frac{Q}{B}\right)_j DC_{ij} + Y_i + E_i + BA_i + B_i \left(\frac{P}{B}\right)_i (1 - EE_i) \text{ (Eq. 1)}$$

The second equation (Eq. 2) ensures energy balance, calculating consumption of the ith group (Q) as the sum of its production, respiration (R), and excretion (U)

$$Q_i = P_i + R_i + U_i \qquad \text{(Eq. 2)}$$

Functional groups have been chosen to be the same on both models for the Atlantic and the English Channel. Thus, the three models developed in this study were composed of 23 functional groups or functional compartments ranging from detritus to seabirds and mammals. Seabirds were divided into two compartments, according to their feeding strategies. The "plunge and pursuit divers" compartments were mainly composed of gannets and the "surface feeders" were mainly composed of gulls and kittiwakes. Marine mammals (*Delphinus delphis, Stenella coeruleoalba and Tursiops truncatus*) were gathered in one compartment. Cephalopods were divided into two compartments: the benthopelagic ones mainly composed of *Loligo vulgaris* and the benthos ones mainly composed by *Sepia officinalis*. The model was also composed of seven compartments of fish (gadidae, piscivorous, benthos feeders, Labridae, Sparidae, flatfish and planktivorous). Gadidae, Labridae and Sparidae were not aggregated with the other compartment in order to see in details the potential impact of the reef effect on these three compartments. Benthic invertebrates were divided into six compartments (commercial decapod, predators, scavengers/omnivores, filter feeders, surface deposit feeders and subsurface selective feeders). Finally, the model was also composed of one compartment of zooplankton, one compartments of bacteria, one compartments of phytoplankton and compartments group of detritus. For all these functional compartments data were collected from the literature and/or from field measurement. For instance, for the “BAR_landes” model, the biomass of fishes and benthos species have been collected from bottom-trawl survey conducted by the EVHOE cruise, IFREMER since 1997 in the Bay of Biscay. However, for the “AR_landes” model, the fish biomass was estimated from under water visual census campaingns (the survey covered a total surface of 102m² corresponding to the footprint of the three types of ARs) and benthos biomass were estimated from benthos samples which were collected on summer 2019 and winter 2020. Six scraps, on different ARs position (inside, below, North, East, West and South) were made by divers on each ARs type, using a quadrat sampler of 20*20cm for 2019 and 30*30cm for 2020, a putty knife and a net bag (<1mm mesh size). A total of 0.553m² were sample on summer 2019 and 0.819 m² on winter 2020.

In addition, this EwE approach also allows providing measure of the ecosystem emergent properties through the calculation of Ecological Network Analysis indices (ENA). ENA is one well- known method to quantify how species interact with and influence their environment [10]. For instance, these indices enable the characterization of the flow recycling, redundancy and omnivory, or the trophic levels of each functional group [11]. Such indices have been recently proposed as ecosystem health indicators for describing the food web functioning in different contexts, including the implementation of the Marine Strategy Framework Directive in Europe [12].

3. Results

For the moment, only the model representing the present initial state of the ecosystems, depicting the ecosystem before the installation of AR in the Rade of Cherbourg and in the Capbreton as well as after the installation of the three artificial reefs of the Capbreton concession, have been analysed. Thus, here, we only present the first results based the comparison of the BAR_Landes and the AR_Landes models (Table 1 ; Figure 4).

Results showed that the total living biomass was higher after the installation of the ARs. In fact, the total living biomass increased by approximately 28% after the installation of the three ARs of the Capbreton concession. This increase was mostly due to the increase in Labridae and fish benthic feeders, flatfish which showed an increase by a factor of approximately 10, 7, 3 respectively, after the installation of ARs.

In the both Landes models (BAR and AR), trophic levels (TL) of the functional groups ranged from 1 for primary producers and detritus to a maximum of 4.8 for marine mammals that can be thus considered as top predators in the area (Table 1; Figure 4). As mentioned below, TL 1 was composed of two compartments (primary producers and detritus, as imposed by the model construction) and represented approximately 54 % and 46 % of the total biomass in the BAR and AR model respectively. TL 2 was composed of six functional compartments (Bacteria, zooplankton, benthic invertebrate subsurface and surface deposit feeders and filter feeders) making up approximately 24% and 32.5% of the total biomass in the BAR and AR model respectively. TL 3 encompassed the fish functional group major part (such as flatfish, benthos feeders, planktivorous, Sparidae) and it was composed of nine functional compartments. It represented 18 and 17% of the total biomass in the BAR and AR model respectively. Finally, the TL 4, was composed of five functional compartments in the BAR model and of six functional compartments in the AR model. TL 4 which corresponded to top predators represented only 3 and 4% of the total biomass in the BAR and AR model respectively. Thus, the TL contributing most to the total biomass in the both models was the TL 2.

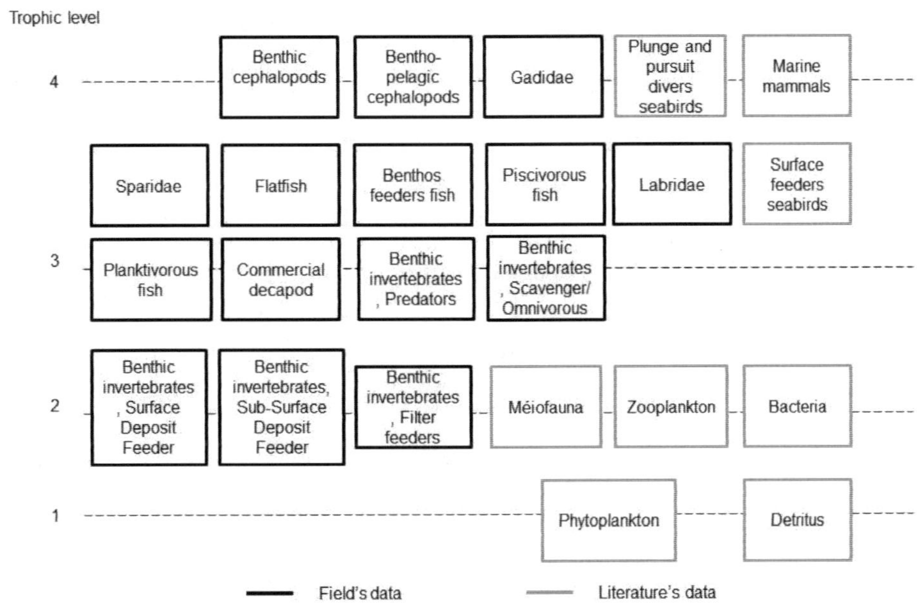

Figure 4. Trophic modelling of the Capbreton before installation of Artificial Reefs (BAR_Landes model).

Results based on the ENA indices analyses revealed a combination of changes in the ecosystem structure and functioning after the installation of the three ARs of the Capbreton concession. After the installation of the three ARs of the Capbreton concession, the maturity of the ecosystem tends to increase (according to ecological theories developed by Odum, 1969, 1971 [13, 14], and Ulanowicz, 1986 [15]).

Table 1. Biomass values (gC.m^{-2}) and Trophic Levels, in the two Ecopath models ("before" (BAR_Landes) and "after" (AR Landes) the construction of the artificial reef).

	Biomass		Trophic Level	
Functionnal groups	BAR model	AR model	BAR model	AR model
Plunge and pursuit divers seabirds	0.0001	0.0001	4.06	4.11
Surface feeders seabirds	0.0001	0.0001	3.93	3.94
Marine mammals	0.0018	0.0018	4.85	4.83
Benthopelagic cephalopods	0.0096	0.0037	4.31	4.17
Benthic cephalopods	0.0240	0.0371	4.10	4.06
Gadidae	0.0487	0.0409	4.03	4.00
Fish, piscivorous	0.1758	0.3065	3.98	4.20
Fish, benthos feeders	0.0246	0.1705	3.78	3.43
Labridae	$3,9 \times 10^{-6}$	4×10^{-5}	3.51	3.31
Sparidae	0.0263	0.0089	3.56	3.62
Fish, flatfish	0.0208	0.0672	3.53	3.36
Fish, planktivorous	0.2231	0.2670	3.15	3.15
Commercial Decapod	0.0322	0.1543	3.13	3.12
Benthic invertebrates, Predators	0.2920	0.6161	3.16	3.22
Benthic invertebrates, Scv/O	0.8228	0.2827	3.14	3.20
Benthic invertebrates, Filter feeders	0.3073	1.2196	2.28	2.32
Benthic invertebrates, sDF	0.2690	0.3366	2.44	2.36
Benthic invertebrates, ssDF	0.2883	0.4126	2.22	2.22
Meiofauna	0.2642	0.2642	2.29	2.21
Zooplankton	0.3600	0.3600	2.15	2.15
Bacteria	0.3940	0.3940	2.02	2.02
Phytoplancton	1.3800	1.3800	1	1
Detritus	2.8467	2.8467	1	1

4. Discussion

Trophic network modelling has been developed over decades and has been applied to various marine ecosystems around the world. This approach has been particularly used to understand the effect of fisheries on the entire ecosystem [5]. Recently, this approach has been extended to other research domains such as management of Marine protected Area [16] or to simulate the effect of wind farming on the ecosystem [9, 17]. Like Offshore Wind Farms (OWF), ARs are mostly deployed on soft bottom habitats. They create hard substrate that will be colonized by different populations. One of the main results of this study was the increase in ecosystem maturity after the installation of the three ARs of the Capbreton concession which is in accordance with the results of Raoux *et al.,* 2017, 2019 [9, 18] and Wang et al., 2020 [19]. In fact, Raoux et al. (2017) [5] analysed the potential effect of OWFs and more particularly reef effects on ecosystem energy flows by describing the pre- and post-construction state of the Courseulles-sur-mer ecosystem using biological data and hypothetical scenarios after 30 years simulated by Ecosim. The results also indicated an increase in ecosystem maturity. However, these studies were based solely on simulated future scenarios instead of using actual post-construction field investigations. Finally, Wang et al. (2019) [19], also built an Ecopath model applying it before and after the construction of the Rudong OWF (Jiangsu coastal ecosaystem, China) based on biological field data collected before and after the construction of the OWF. These results also indicated that the ecosystem tended to develop towards higher maturity following the OWF construction with the reef effect. ENA comparison between the model "Before Artificial Reef _Cherbourg" and «After artificial Reef_Cherbourg » could also confirm this trend. In this study, using ENA on this AR ecosystem makes it possible to highlight the trophic modification linked to the introduction of hard substrate on soft habitats. Thus, these results indicate the interest of using a large set of ENA indices to characterise different trophic functioning attributes, essential for a complete view of the induced changes [18].

Acknowledgements
This work was co-funded by National Association for Technical Research and Atlantique Landes Récifs (ALR). We also acknowledge, for their help in compiling the datasets and for giving expert advices: Eric Charbonnel, Philippe Lenfant, Nelly Ferrou-Rocher, Andréa Bayens and all ALR's volunteers

References

[1] Baine M. 2001, Artificial reefs: a review of their design, application, management and performance. *Ocean Coast. Manag*. **44**, 241-259.
[2] Lee MO, Otakeb S, Kima JK. 2018, Transition of artificial reefs (ARs) research and its Prospects. *Ocean Coast. Manag*. **154**, 55-65.
[3] Tessier A, Bodilis P, Charbonnel E, Dalias N, Francour P, Lenfant P, Seaman W. 2015, Assessment of French artificial reefs: due to limitations of research, trends may be misleading. *Hydrobiologia* **753**, 1-29.
[4] Fabi G, Bellan-Santini D, Charbonnel E, Çiçek B, Goutayer García JJ, Jensen AC, Kallianiotis A, Neves dos Santos M, Spagnolo A. 2011, Overview on artificial reefs in Europe. Braz. *J. Oceanogr* **59**, 155-166.
[5] Salaun J, Pioch S, Dauvin JC. 2020, Description of a theoretical social-ecological approach to manage artificial reefs. Vie Milieu.
[6] Colléter M, Valls A, Guitton J, Gascuel D, Pauly P, Christensen V. 2015, Global overview of the applications of the Ecopath with Ecosim modelling approach using the EcoBase models repository. *Ecol. Model*. **302**, 42–53.
[7] Heymans JJ, Coll M, Link JS, Mackinson S, Steenbeek J, Walters C, Christensen V. 2016, Best practice in Ecopath with Ecosim food-web models for ecosystem-based management. *Ecol Model*. **331**, 123–128.
[8] Christensen V, Walters CJ. 2004, Ecopath with Ecosim: methods, capabilities and limitations. *Ecol. Model*. **172**, 109–139.
[9] Raoux A, Tecchio S, Pezy JP, Degraer S, Wilhelmsson D, Cachera M, Ernande B, Lassalle G, Leguen C, Grangeré K, Le Loc'h F, Dauvin JC, Niquil N., 2017, Benthic and fish aggregation inside an offshore wind farm: Which effects on the trophic web functioning? *Ecol. Indic*. **72**, 33–46.
[10] Haak DM, Fath BD, Forbes VE, Martin DR, Pope KL. 2017, Coupling ecological and social network models to assess "transmission" and "contagion" of an aquatic invasive species. *J. Environ. Manag*. **190,** 243-251.
[11] Latham LG. 2006, Network flow analysis algorithms. *Ecol. Model*. **192**, 586–600.
[12] Niquil N, Le Loc'h F, Tecchio S, Chaalali A, Vouriot P, Mialet B, Fizzala X, Féral JP, Lamare S, Dauvin JC, Safi G. 2014, "Ongoing research on ecosystem health indicators for food webs in the MSFD context", Trans-Channel forum proceedings "Science and Governance of the Channel Marine Ecosystem" Caen, France, 14-15.
[13] Odum EP. 1969, "The strategy of ecosystem development". Science **164**, 262-270.
[14] Odum EP. 1971, "Fundamentals of Ecology", WB Saunders Co., Philadelphia, USA. 574 pp.
[15] Ulanowicz RE. 1986, "Growth and Development: Ecosystems Phenomenology". *Springer-Verlag,* New York. 166 pp.
[16] Hermosillo-Núñez B, Cupul-Magaña AL, Ortiz M, Rodríguez- Zaragoza FA. 2018, Trophic network properties of coral ecosystems in three marine protected areas along the Mexican Pacific coast: assessment of systemic structure and health. *Ecol. Complex*. **36**, 73-85.

[17] Pezy JP, Raoux A, Dauvin JC. 2019, An ecosystem approach for studying the impact of offshore wind farms: a French case study. *ICES J. Mar. Sci.* **125**, 1–9.
[18] Raoux A, Lassalle G, Pezy JP, Tecchio S, Safi G, Ernande B, Mazé C, Lequesne J, Girardin V, Dauvin JC, Niquil N. 2019, Measuring sensitivity of two OSPAR indicators for a coastal food web model under offshore wind farm construction. *Ecol. Indic.* **96**, 728–738.
[19] Wang J, Zou X, Yu W, Zhang D, Wang T. 2019, Effects of established offshore wind farms on energy flow of coastal ecosystems: a case study of the Rudong offshore wind farms in China. *Ocean Coast. Manag.* **171**, 111– 118.

IOP Conf. Series: Materials Science and Engineering 1245 (2022) 012005 doi:10.1088/1757-899X/1245/1/012005

Effectiveness of ecomodules in increasing aquatic biodiversity and bioproductivity in a port environment.

P Paalvast[1], R Wolf[2] and M Faasse[3]

[1] ecoconsult, Esther Vergeerstraat 31, 3118 JK Schiedam, the Netherlands
[2] Port of Rotterdam Authority, PO Box 6022, 3002 AP, the Netherlands
[3] Eurofins Aquasense, Korringaweg 7, 4401 NT, Yerseke, the Netherlands

Corresponding author: peter.paalvast@gmail.com

Abstract. In early spring of 2018, so called ecomodules were attached to support piles of jetties within the polyhaline part of the port of Rotterdam. The objective of the ecomodules is to increase bioproductivity and biodiversity of the port basins and waterways, by 'greening' nautical infrastructure structures, which would otherwise support few species. An ecomodule consists of a coated steel construction of 1 meter wide and 4 meters high to which elements of structural diversity that facilitate the settlement and growth of organisms can be attached. About a quarter of the ecomodule protrudes above the water level at low tide, the other part is permanently under water. Five different types (I to V) of ecomodules have been tested. Ecomodules with: (I, cups, reference) tidal cups and a subtidal concrete reference slab, (II, ropes) used nylon hawsers with a diameter of 12 cm, (III, drainage pipes) drainage pipes sheathed with polypropylene with a diameter of 12,5 cm, (IV, fender hulas) bundles of thin nylon threads (untangled used hawsers), and (V roots) strands with 3 mm thick nylon strings with a length of 40 cm each. Monitoring of the ecomodules took place in the summers of 2018 and 2019.

The biofouling on all ecomodules was zonate both inter- and subtidal. Different types of seaweed were found of which Ulva spec. occurred on all ecomodules while the large seaweed Fucus vesiculosus (Bladder wrack) only occurred on type II and IV. Amongst the macrofauna some groups/species clearly (co-)dominated the different types of ecomodules: barnacles and *Crassostrea gigas* (Pacific oysters) on type I, ascidians on type I, II, III and IV, hydrozoans on type II, III and V, bryozoans on type IV, while with a coverage of nearly 100% Mytilus edulis (Blue mussel) dominated type V.

The ecomodules with elements of structural diversity vs the concrete reference plate scored higher on both biodiversity and bioproductivity. The total number of macrofaunal species found on the ecomodules ranged from 23 to 39 in order from lowest to highest number on respectively type, I (concrete reference plate), IV, V, II, III. The wet biomass on the ecomodules ranged from 5225 gr.m-2 to 80096 gr.m-2 in order from lowest to highest weight on respectively type, I (concrete reference plate), IV, III, II, V.

We conclude that ecomodules can be an effective means to increase biodiversity and biomass in an aquatic environment consisting to a large extent of artificial structures, such as a port. An ideal ecomodule was designed based on the findings of the experiment and will be applied to mooring posts in a new project.

1. Introduction

The port of Rotterdam (Figure 1) is situated in the remains of the Rhine-Meuse estuary. It has a total surface of 12.603 ha of which 4.810 ha is water area. The total length of the Port of Rotterdam is 42 km and covers the complete salinity gradient from fresh to seawater. Rotterdam city is located about 35 km upstream from the sea. It has developed from a small fishing port in the 14th century to the largest port of Europe nowadays. Large harbour expansions took place in particular three periods: a) so-called Golden Age of the Netherlands (1600-1700), b) the industrial revolution in the mid-19th century and c) after the second World War (1945-nowadays) in particular [1]. The development of the port of Rotterdam transformed from a formerly soft sediment habitat into a habitat that is mainly dominated by hard structures such as retaining walls, rip rap, piers, quays, pontoons, and mooring poles of concrete, hardwood and steel. As an example in 1835 the port of Rotterdam area comprised 4800 ha soft and 16 ha hard intertidal area and by 2008 this had changed to 17 ha and 381 ha respectively [2]. In the large polyhaline harbours of Europoort, Maasvlakte and Maasvlakte 2 these hard structures were quickly colonised by hard substrate communities new to area.

Before the construction of Maasvlakte 2 (on 2000 ha of reclaimed land between 2008-2013), there was little interest from the authorities in the consequences of port development on estuarine nature. But nowadays within new projects measures are taken to improve the quality of the port's aquatic ecosystem dealing by greening nautical infrastructure.

In early spring of 2018 so called ecomodules were attached to new support piles of jetties within the polyhaline part of the port of Rotterdam. The aim was to test whether ecomodules covered with different structures, such as ropes and pipes, locally increase biodiversity and bioproductivity. Where possible, discarded materials have been used.

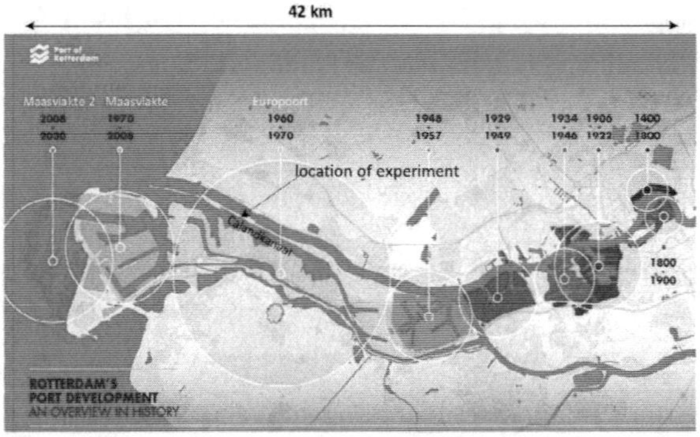

Figure 1. Overview of the port of Rotterdam area, its historical development and the location of the experiment.

2. Material and methods

2.1. Location

The colonisation experiment with the ecomodules was carried out in the Calandkanaal close to the north bank (Fig. 1) and 16 ecomodules were attached to 8 support piles of newly constructed jetties in the third week of April 2018. The depth of the water at that location is between 10 and 11 m below the MLWL (mean low water level). It's a relative lee environment. Waves, even in big storms, and tidal currents are moderate. However, there is 24-h activity of vessels and their propellers produce strong water currents, often resulting in sediment resuspension. The sediment mainly consists of sand, so that it quickly sinks to the bottom again. The average tidal range is 1.94 m. At springtide the tidal cycle exhibits approximately a 4 h period of flood, a 4 h period of ebb and a 4.5 h low water period. The water is often clear with a Secchi depth up to 3 m. The water is always polyhaline and depending on the river discharge there is a more or less clear vertical gradient in salinity with high salinity near the sea floor and a lower salinity at the water surface.

2.2. Ecomodules

The basis of an ecomodule is formed by a steel construction of 1 meter wide and 4 meters high to which the different structures are attached. With the help of a crane, the ecomodules can be hung on steel straps that are attached to the support piles of the jetties (Fig. 2) or lifted for inspection and monitoring.
Five different types (I to V, Fig. 3) of ecomodules have been tested. These ecomodules were attached on the east side of a support pile.

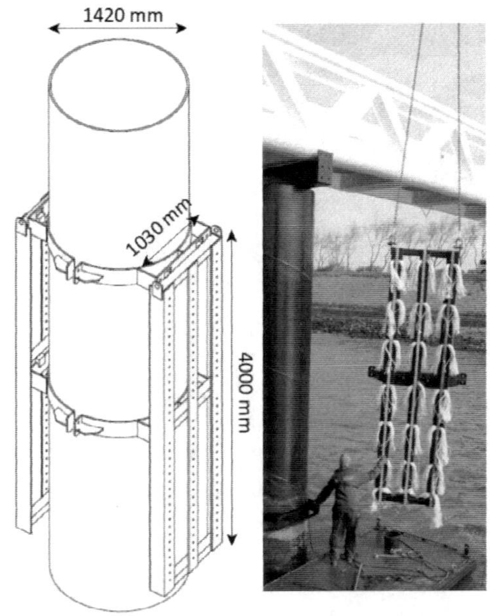

Figure 2. Design drawing of the basis of an ecomodule (left) and the mounting of type IV on a support pile of a jetty (right).

2.2.1. Type I, ecomodule reference. The ecomodule reference comprises a concrete slab and four tidal cups. The concrete slab serves as a reference for the part of the other ecomodules below mean low water level (MLWL). The dimensions are 2.5 m high and 1.03 m wide, making the surface approximately 2.5 m^2.

The tidal cup part is in fact a separate module. It includes 4 galvanized sinks measuring 55 cm x 24 cm x 35 cm (width x depth x height) and these are evenly distributed between MLWL (-71 cm NAP, NAP=Amsterdam Ordnance Datum) and the top edge of the module (+ 79 cm NAP). In the design, the sinks are provided with a network to stimulate algae growth. For practical reasons, this network was not replaced after the first monitoring. Each sink contains a building block measuring 35 cm x 20 cm x 20 cm (length x width x height). The building blocks are made of raw concrete and provided with cavities and holes.

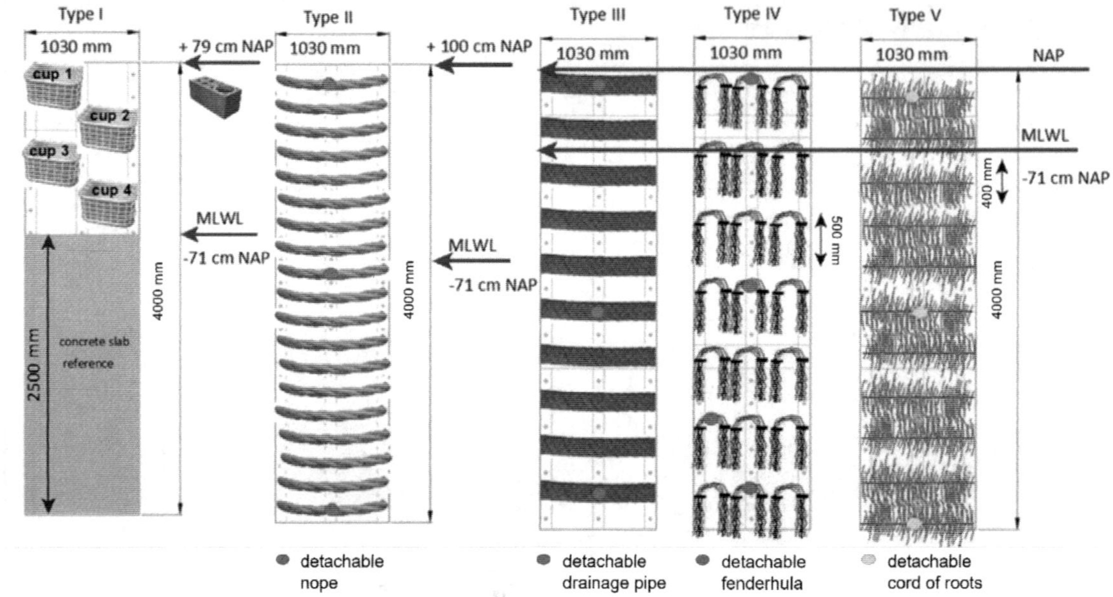

Figure 3. Design drawings of the ecomodules and their position in relation to NAP. NAP = Amsterdam Ordnance Datum.

2.2.2. Type II, ecomodule ropes. The ecomodule 'ropes' is made up of a zigzagging rope (or hawser) of which the loops are evenly distributed over the ecomodule and 3 detachable horizontal ropes. Of the detachable ropes, one is located above the loops, one below and one in the middle. The diameter of the ropes is 125 mm. The top of the ecomodule is at +100 cm NAP. 8 ropes are above MLWL.

2.2.3. Type III, ecomodule drainage pipes. On the ecomodule 'drainage pipes' nine pieces of drainage pipe wrapped in polypropylene fibre are mounted. Although they are called the drainage pipes, they do not have any pinholes. As a result, the top two pipes remain partially filled with water above the low water level at low tide. Three of the pipes are detachable, namely the numbers 1, 6 and 10 seen from above.

2.2.4. Type IV, ecomodule fender hulas. The ecomodule 'fender' hulas has 9 rows, each with 3 bundles of very thin nylon wires (untangled used hawsers) with a length of 50 cm. One row is located above MLWL. One bundle above and three below MLWL are detachable.

2.2.5. Type V, ecomodule roots. An element of ecomodule 'roots' (named a cord of roots) consists of two polypropylene lashing straps sewn together, 2 cm wide with polypropylene strands between 30 cm and 40 cm in length and a thickness of 3 mm. One of the cord of roots of the polypropylene had been replaced by coconut twine. The elements were provided by SEABOOST, Montpellier, and have proofed to be successful in France [3].

2.3. Monitoring
The monitoring of the ecomodules (5 out of 16) took place on 27/28 August 2018 (T1) and 17/18 June 2019 (T2). Respectively 19 and 61 weeks after placement in the third week of April 2018.

2.3.1. Abiotics. On the monitoring dates samples of the surface water and water at a depth of -3.85 NAP were taken and oxygen concentrations were measured with a HQ20 Hach Portable LDOTM Dissolved Oxygen/pH Meter. Salinity and temperature were measured with a WTW-conductivity meter (Cond

3301) with a TetraCon®325 conductivity and temperature sensor. Water transparency was measured with a standard Secchi disc.

2.3.2. Biotics. The biotic parameters comprised, the length of zonation of algae and macrofauna, % coverage of algal species, number of algal species, % coverage or estimated number of individuals of sessile macrofauna species, number of sessile macrofauna species, estimated number of individuals of mobile macrofauna species, number of mobile macrofauna species. The detachable elements were rinsed in a container and the rinse water was poured over a sieve with a mesh size of 0,5 mm to collect the small mobile species as well. These have been preserved and determined at the laboratory of eCoast Marine Research at Yerseke, the Netherlands. For the abundance of species (the degree of presence), the coverage percentages of algae, oysters, mussels, colonial ascidians, encrusting bryozoans, barnacles, hydroids, tube-dwelling polychaetes and sponges and the numbers of sessile and mobile solitary species per m^2 have been converted into a coding according to a (adapted) scale developed by van der Maarel [4](table 1). The use of abundance scale categories is a standard procedure in intertidal monitoring of rocky coasts [5].

The wet biomass on a detachable element of an ecomodule was determined by weighing it with an accuracy of 1 gram at the monitoring moments and subtracting the weight of the element itself in wet condition. The latter has been used to calculate the biomass per square meter per ecomodule. From the reference plate, a representative surface of 416 cm^2 has been stripped of all biofouling in order to calculate the biomass per m^2.

Table 1. Abundance scale code for seaweeds and macrofauna.

code	percentage of cover	number of specimen
1	0% - 0.01%	1
2	0.01% - 0.1%	2 – 10
3	0.1% - 1.0%	11 – 100
4	1% - 5%	101 – 500
5	5% - 12.5%	>500
6	12.5% - 25%	
7	25% - 50%	
8	50% - 75%	
9	> 75%	

3. Monitoring results.

3.1. Abiotics

In the period between the installation of the ecomodules and the second monitoring moment, no extremes in the water temperature occurred, despite the warm summers. The salinity of the water was significantly higher at T1 in 2018 than in 2019 at T2 (Table 1). This was due to the prolonged low river discharge between the beginning of July and the end of November in 2018. The oxygen concentration was almost at the saturation level. Since 2003, regular oxygen measurements have taken place in the Calandkanaal. Oxygen concentrations were always above 90%. The secchi depth of the water in the Calandkanaal is generally good. At both monitoring moments during low tide, the visibility was approximately 3 meters. This means that every now and then light must have reached the lower end of the modules.

Table 2. Abiotic conditions at the monitoring data.

date	**26-8-2018**	**17-6-2019**
time	11:35	11:18
Secchi depth (m)	2.9	3.2
surface		
salinity	25.3	20.5
temperature (°C.)	19.1	18.6
oxygen saturation (%)	93.1	95.9
oxygen (mg/l)	7.5	7.5
3.5 m below surface		
salinity	25.7	21.6
temperature (°C.)	19.4	17.6
oxygen saturation (%)	96,9	95.6
oxygen (mg/l)	7.7	7.7

3.2. Ecomodules

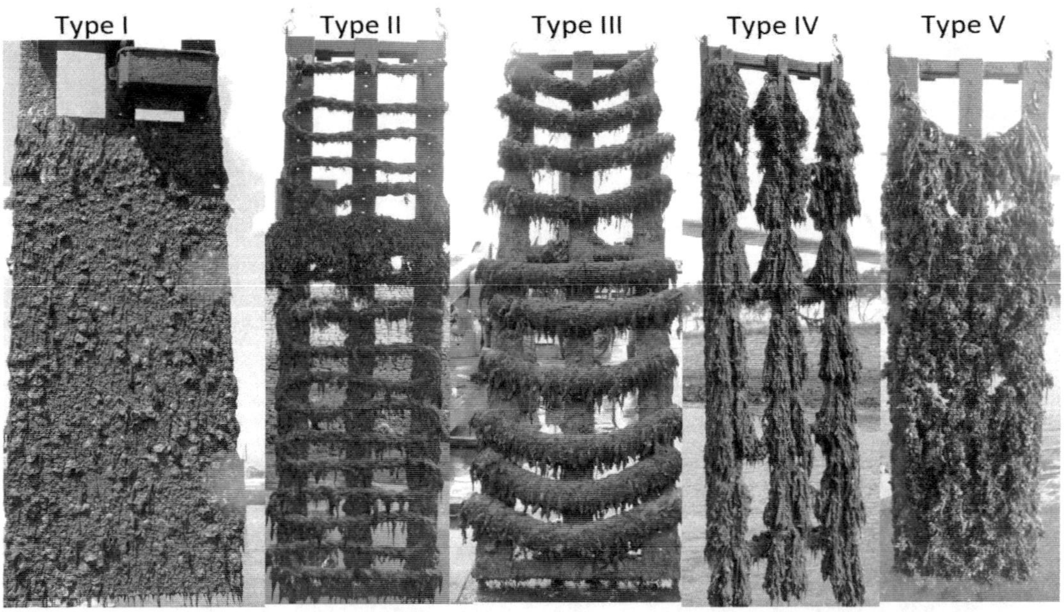

Figure 4. The ecomodules and their biofouling on T2.

Table 3. Abundance of seaweeds and macrofauna above MLWL on the ecomodules and reference pile. C= tidal cup. *position relative to NAP in cm. Z=Zone. "length of zone in cm. # non-native species.

Group	Species	Type I								Type II						Type III		Type IV		Type V		ref pile	
		T1				T2				T1			T2			T1	T2	T1	T2	T1	T2	T1	T2
		C1	C2	C3	C4	C1	C2	C3	C4	Z1	Z2	Z3	Z1	Z2	Z3	Z	Z	Z	Z	Z	Z	Z	Z
		+79*	+44*	+9*	-26*	+79*	+44*	+9*	-26*	80"	20"	50"	20"	40"	80"	80"	80"	70"	70"	60"	60"	80"	80"
Chloropyceae	*Blialingia minima*									8			9									6	6
	U. intestinalis/prolifera/B. minima																5		5				
	Ulva intestinalis	4	4	4	4						7	6											
	Ulva intestinalis/prolifera					8	7	4	6						5	7							6
	Ulva rigida					4	4	6	4			3		3	3	7	7	7	6	7	9		
Rhodophyceae	*Aglaothamnion hookeri*											3					4						
	Porphyra purpurea									3	3		3	3		4	2			4	3		
Phaeophyceae	*Fucus vesiculosus*					4	6	6	6		4	6		4	8	3	5						
Amphipoda	*Microdeutopus gryllotalpa*								3														
	Monocorophium insidiosum								3														
	Ptilohyale littoralis #			2																			
Ascidiacea	*Ascidiella aspersa*								1							3							
	Botrylloides violaceus #			2	2							4				3							
	Botryllus schlosseri	2	3	2	2			2	2								2						
	Molgula manhattensis #		1	2	2			2	2							5		3					
	Styela clava #																2						
Bivalvia	*Crassostrea gigas #*						2	2	2					1	2		2						
	Mytilus edulis				1										3	3	2		2	3	3		
Bryozoa	*Bowerbankia sp*						1										2						
	Conopeum reticulum								2								1						
	Conopeum seurati		2		2																		
	Cryptosula pallasiana			2			2	2	2								2						
	Electra pilosa																2						
	Tricellaria inopinata #				2							4				3		5	2				
Cirripedia	*Amphibalanus improvisus*	2	7	4	7	4	5	4	5							3	2	3					
	Austrominius modestus #					2										3	3		2	3	3	5	8
	Balanus crenatus															3							
Decapoda	*Brachyura sp.*								3														
	Carcinus maenas	2	2	2	3				2							2					3		
	Hemigrapsus takanoi #														2								
	Palaemon elegans							1															
Hydrozoa	*Gonothyraea loveni*						2										6		3				
	Clytia hemisphaerica															3		4					
	Obelia dichotoma															3							
	Opercularella lacerata			2	2			3	4														
Pisces	*Pholis gunnellus*								2														
Polychaeta	*Alitta succinea*						1																
Serpulidae	*Ficopomatus enigmaticus #*	2	1	2	2		2	2	2														
Porifera	*Halichondria panicea*			2	3	2		2	2														
Total number of species		5	8	12	13	7	11	13	20	2	3	6	2	4	7	17	17	6	6	4	5	2	3

It is unknown how many mobile animal species that were on, in or near the ecomodules fled or fallen off the module during lifting. When taking it up and on board, large quantities of starfish, common shore crabs and butterfish fell from Type IV and V in particular.

Table 4. Abundance of species sampled by rinsing of the detachable elements of the ecomodules above MLWL. # non-native species

Group	Species	Type II		Type III		Type IV	
		T1	T2	T1	T2	T1	T2
Amphipoda	*Caprella mutica #*			2			
	Gammarus locusta					1	2
	Jassa marmorata		1	2	2		
	Microdeutopus gryllotalpa		2	2	2	1	
	Monocorophium insidiosum				3		
	Monocorophium spec.			2			
	Ptilohyale littoralis #					2	
	Stenothoe marina			3			
Decapoda	*Carcinus maenas larvea*				3		
Hexapoda	*Aphrosylus celtiber*		1				
	Diptera sp.		3				
Entoprocta	*Barentsia spec.*			1			
	Pedicellina sp.				1		
Gastropoda	*Tergipes tergipes*			1			1
Polychaeta	*Alitta succinea*				1		
	Autolytus prolifer			1			
	Mysta picta		2				
	Polydora cornuta		2				
	Proceraea cornuta				3		3
	Syllidia armata		3	1			
Total number of species		0	7	9	7	3	3

3.2.1. Biodiversity. Figure 4 shows the biofouling on the various ecomodules at T2. The differences are remarkable. Above MLWL (Table 3 and 4) the most common seaweed to be found and in the highest densities was sea lettuce (Ulva rigida) that developed on all eco modules except on the reference pile. The other green seaweeds were often a mixture of species. Red seaweeds occurred in rather low densities and were absent on Type I, IV and the reference pile. Only one species of brown seaweed occurred, bladderwrack (Fucus vesiculosus) and was best developed on type II. The biofouling on Type II showed a clear zonation as its section above MLWL was around 150 cm. The macrofauna was best represented in the Type I tidal cups and the Type III drainage pipes, both of which retained water during low tide.
Table 5 shows the abundance (degree of presence) of the fouling species below MLWL on the ecomodules and table 6 the abundance of species sampled by rinsing the detachable elements. Except on Type I (on T1) and Type V there was a clear zonation due to the presence and absence of seaweeds. Among the seaweeds *U. rigida* was mainly dominant in Z1 (Zone 1) and *Agloathamnion hookeri* in Z2.

Table 5. Abundance of seaweeds and macrofauna below MLWL on the ecomodules. Z=Zone. "length of zone in cm. # non-native species. *only dead individuals.

Group	Species	Type I			Type II				Type III				Type IV				Type V	
		T1	T2		T1		T2		T1		T2		T1		T2		T1	T2
		Z	Z1	Z2	Z1	Z2	Z1	Z2	Z1	Z2	Z1	Z2	Z1	Z2	Z1	Z2	Z	Z
		250"	40"	210"	150"	50"	150"	50"	120"	200"	120"	200"	190"	140"	190"	140"	340"	340"
Chlorophyceae	*Bryopsis plumosa*	1					1											
	Ulva rigida	6	7		5	2	5	3	5		5		3		5		3	3
Rhodophyceae	*Aglaothamnion hookeri*	4		5	4	4	5	5	4		6	6					3	4
	Ceramium tenuicorne												4		3			
	Porphyra purpurea		5															
Ascidiacea	*Ascidella aspersa*					3	3	3	3	3				3		3	3	
	Botrylloides violaceus #	6		3	4	5	4	5		3		3		3		3	3	3
	Botryllus schlosseri			7	3		3		2	2	4	3	3	3	3	3	3	
	Ciona intestinalis									3								
	Molgula manhattensis #				3		2		5	5		3	5	5	3	3	4	3
	Styela clava #	1			1		3	3	2	2	2	3		1		3		3
Asteroidea	*Asterias rubens*									3		2	2	3	2	3	3	3
Bivalvia	*Crassostrea gigas #*		6	7				2				2						
	Mytilus edulis	2			3	3	3		2		2	2	5	4	3	3	9	9
Bryozoa	*Alcyonidium mytili*																	2
	Bowerbankia sp.						3					2				3		
	Conopeum reticulum															1		2
	Conopeum seurati				1		1											
	Cryptosula pallsiana	2					3					2						
	Electra pilosa			1			3	3										
	Tricellaria inopinata #	4		3	4	4	4	3	3	5		2	6		3	3	3	3
Cirripedia	*Amphibalamus improvisus*	4/9*		7			3		3	2	2	2		3		3	3	3
	Austrominius modestus #		9		2													
	Balanus crenatus	2				3			2					2			3	
	Semibalanus balanoides		2															
Decapoda	*Cancer pagurus*									1								
	Carcinus maenas				3		3	3	3	3		4	4	3	2	4	5	4
Hydrozoa	*Clytia hemisphaerica*									3								
	Ectopleura larynx																	5
	Gonothyraea loveni		2	6							7	8			3	5		
	Obelia dichotoma	2			3	3	6	6	3	3			3	3			3	
Pisces	*Pholis gunnelus*							3				4	2	3	2	4	4	4
	Gobiusculus flavescens													1				
	Agonus cataphractus							1										
	Myoxocephalus scorpius							1		1								
Serpulidae	*Ficopomatus enigmaticus #*	3		2								2						
	Hydroides ezoensis #	2		2														
	Neodexiospira brasiliensis #			3														
Porifera	*Halichondria panicea*	1		2				3		2		2	3		3	3		3
	Haliclona oculata										2	2						
Total number of species		13	6	12	12	8	17	14	12	15	8	18	11	13	11	15	14	15

IOP Conf. Series: Materials Science and Engineering 1245 (2022) 012005 doi:10.1088/1757-899X/1245/1/012005

Table 6. Abundance of species sampled by rinsing of the detachable elements of the ecomodules below MLWL. # non-native species.

Group	Species	Type I		Type II		Type III		Type IV		Type V	
		T1	T2	T1	T2	T1	T2	T1	T2	T1	T2
Amphipoda	*Apohyale prevostii*							1			
	Caprella mutica #	1		3		3		3		3	
	Gammarus locusta			2	2				1		1
	Jassa marmorata			3	2	2				2	
	Melita nitida #	2						1		1	
	Melita palmata	2		3	3	3		4	4	3	5
	Microdeutopus gryllotalpa	2	3	4	5	3	3	4	5	3	5
	Monocorophium acherusicum			3		3	1	2			
	Monocorophium insidiosum		2		4		3		3		3
	Monocorophium sextonae #	2		3							
	Monocorophium spec.									1	
	Stenothoe marina	1		3		2		1		2	2
	Stenothoe valida										2
Bivalva	*Abra tenuis*						2				
	Ruditapes philippinarum #			2			1				
	Theora lubrica #						2				
Decapoda	*Brachyura*		3								
	Carcinus maenas juveniles								4		
	Hemigrapsus takanoi juv. #					2	1			1	1
	Palaemon elegans		1								
	Palaemon macrodactylus #					2	1	2	1		
	Palaemon spec.									1	
	Pilumnus hirtellus					1					
	Pinnotheres pisum										1
Hexapoda	*Diptera sp larvae*		1	2	2		2				
Entoprocta	*Barentsia spec.*					1					
	Pedicellina cernua					2					
	Pedicellina sp.						2		1		
Mysida	*Hemimysis lamornae*					2					
Nemertea	*Nemertea sp.*	1		1							
Platyhelminthes	*Euplana gracilis #*			2							
	Leptoplana tremellaris			1							
	Stylochus? #		1								
Polychaeta	*Alitta sp.*								1		
	Alitta succinea			3		3	1	1			2
	Alitta virens #						1				
	Autolytus prolifer	1		2		2					
	Capitella capitata						1				
	Eulilia viridis										1
	Harmothoe imbricata					1		2			
	Lanice conchilega					1		1			1
	Mysta picta		1	2	2						2
	Polydora cornuta	2		2		2				2	
	Proceraea cornuta		3				2		2		1
	Scoloplos armiger						1				
	Streblospio spec.					1					
	Syllidia armata	1		3	1	3		2	2	3	
	Tharyx sp.						2				
Total number of species		10	8	18	8	19	16	12	10	11	13

The concrete reference plate of Type I was (on T1 and T2) almost completely covered by barnacles, and to a large extent with colonial ascidians, Pacific oysters (*Crassostrea gigas*), and hydrozoans (T2), tube dwelling polychaetes covered up to 1%. Mobile species (T1/T2) were found only in low numbers.

The biofouling on Type II was characterized by colony-forming, but also solitary ascidians, hydroids and foliose bryozoans. Especially at T1 there was a reasonable number of amphipods and polychaetes. At T2 sponges were clearly present. Three fish species were found.

At T1 and T2, solitary and colony-forming ascidians predominated in the biofouling on Type III. At T2 the hydroid, *Gonothyraea loveni,* covered large parts of the drainage pipes. Many mobile species occurred on and within the drainage pipes. The butterfish, *Pholis gunnellus,* and the common shore crab, *Carcinus maenas,* were present in large numbers.

The biofouling on Type IV was predominated by solitary and colony-forming ascidians, mussels, foliose bryozoans. Large quantities of starfish (*Asterias rubens)*, butterfish and common shore crabs were found on and within the bundles of nylon wires.

The biofouling on Type V was dominated at T1 and T2 by the blue mussel, *Mytilus edulis.* The species occurred in a high density with almost 100% coverage. Dozens of starfish, many hundreds of crabs and butterfish were among the mussels. The rinsed detachable elements contained enormous amounts of amphipods, especially the amphipods *Microdeutopus gryllotalpa* and *Melita palmata.* On T2 the cord of roots with coconut twine had almost completely disintegrated and disappeared.

Table 7 summarizes the number of species found on the elements of all ecomodules. Above MLWL the elements that retain water, Type I tidal cups and Type III drainage pipes, definitely saw more species than the others. Below MLWL Type II and III inhibited the most macrofauna species.

Table 7. Number of seaweed and macrofauna species found on the ecomodules and the wet biomass above (+ MLWL) and below the mean low water level (-MLWL).

position	ecomodule	n seaweed species		n macro fauna species		wet biomass gr.m^{-2}	
		T1	T2	T1	T2	T1	T2
+	reference pile	1	2	1	1	--	--
M	Type I tidal cups	1	3	13	20	--	--
L	Type II	6	6	9	3	2957	1350
W	Type III	4	7	20	19	15300	18300
L	Type IV	2	6	7	8	2700	2357
	Type V	2	2	2	2	167	3717
-	Type I	3	3	24	23	2843	5225
M	Type II	2	3	30	26	17655	67650
L	Type III	2	2	39	34	36233	44550
W	Type IV	2	2	15	25	12205	29877
L	Type V	2	2	23	27	41899	80096

3.2.2. Bioproductivity. Table 7 shows the wet biomass on the different ecomodules at TI and T2. Data concerning the biomass on the reference pile and Type I tidal cups are missing. As far as the former is concerned, it will have been quite low, as there was at T1 a thin layer of *B. minima* and a modest coverage of barnacles only. At T2 a small zone of *Ulva intestinalis* was also present. On Type II, IV and V the biomass production was relatively low compared with Type III of which the elements retained a lot of water during low tide.

Below MLWL the wet biomass production on Type I the concrete reference plate was clearly lower than on the elements of the other ecomodules. Type V scored highest due to the enormous quantities of mussels followed by Type II. Type III and especially Type IV had a lower wet biomass on the elements.

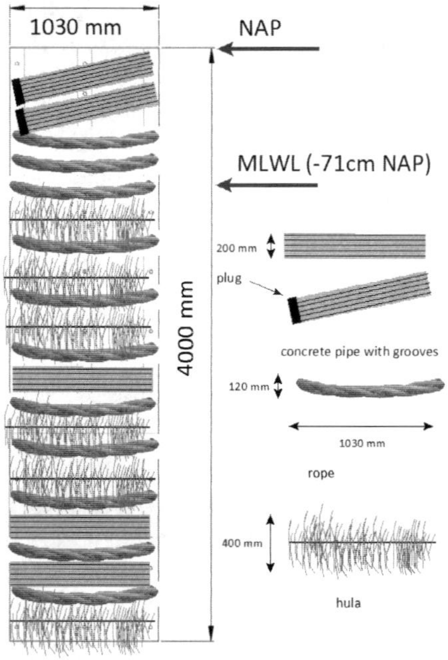

Figure 5. The ideal ecomodule.

4. Discussion

All ecomodules with structural elements scored better on biodiversity than the reference ecomodule, type III the highest, in particular on small mobile species as amphipods and polychaetes. Ascidians were found in high densities on all ecomodules, but de highest densities of solitary species on Type III, IV and V and colonial species on Type I and II. Sponges appeared in approximately equal densities on all ecomodules as did the foliose bryozoan, *Tricellaria inopinata*. The occurrence of the starfish, *A. rubens,* and the common shore crab, *C. maenas,* was probably related to the density of mussels. The more complex the structural element either by itself or created by the fouling community, the more butterfish, *P. gunnellus*, were found. The Pacific oyster, *C. gigas*, except for a few individuals on type II and III, was found exclusively on type I. The polypropylene cords of ecomodule Type V was clearly the substrate on which the mussels settled best. The bioproductivity on Type V was similar to the bioproductivity on pole and pontoon hulas in an experiment carried out 10 years earlier in a small harbour nearby the location of this experiment [9].

A total of 97 macrofaunal species were found, of which 23 were non-native. All but one of these non-native species have been in the Dutch waters and in the port of Rotterdam for quite some time [6, 7]. The exception was the tube dwelling polychaete, *Hydroides ezoensis*, the first finding of the species in the Netherlands [8]. The sea anemone *Metridium senile* was not found on the ecomodules. However the species is found from a depth of 3 meters below MLWL to the bottom on piles on the experimental site. The main reason of its absence was the salinity of the water.

Doubts have been raised about the durability of the polypropylene sheathing of the drainage pipes of Type III, i.e. the risk of the fibers ending up in the environment. The same accounts for the bundles of very thin nylon wires on Type IV, although they are very well attached to the ecomodule. Released fibers could overgrow and be mistaken for food by organisms. This can have disastrous consequences for the organism concerned, for example fish [10, 11, 12]. The advice is not to use them anymore.

An ideal ecomodule (Figure 5) was designed based on the findings of the experiment. In the intertidal zone (+MLWL) from top to bottom, there are two obliquely placed concrete pipes with grooves and three 12 cm thick ropes. The concrete pipes serve as a shelter for prawns and small fish, the grooves on them to facilitate the settlement of algae. The ropes are mainly there as a substrate for large seaweeds. Subtidal there are alternately hulas, ropes and horizontally placed concrete pipes with grooves. The hulas are the substrate for mussels and their associated macrofauna, the ropes mainly for ascidians and hydrozoans and the concrete pipes for barnacles, oysters, et cetera and the inside as a shelter for fish and crabs. This "ideal ecomodule" will be considered for future projects, and applied to mooring posts in a new project elsewhere in the Calandkanaal.

References

[1] Paalvast P 2014 Ecological studies in a man-made estuarine environment, the port of Rotterdam. PhD Thesis. Radboud University Nijmegen the Netherlands

[2] Paalvast P and van de Velde G 2014 Long term anthropogenic changes and ecosystem service consequences in the northern part of the complex Rhine-Meuse estuarine system. *Ocean & Coastal Management* **92** 50-64 https://doi.org/10.1016/j.ocecoaman.2014.02.005

[3] Lapinski M et al 2017 Innovative solutions for the enhancement of biodiversity on coastal infrastructures: case studies of artificial port nurseries and modular mooring reefs. Coastal and Maritime Mediterranean Conference. Edition 4, Split, Croatia https://dx.doi.org/10.5150/cmcm.2017.040

[4] Van der Maarel E 1979 Transformation of cover-abundance values in phytosociology and its effects on community similarity. *Vegetatio* **39** 97-114 https://doi.org/10.1007/BF00052021

[5] Murray S N, Ambrose R F and Dethier M N 2006 *Monitoring rocky shores.* University of California Press. ISBN 0-520-24728-0

[6] Wolff W J 2005 Non-indigenous marine and estuarine species in The Netherlands. *Zoologische mededelingen* **79** 1-116

[7] Gittenberger A et al 2014 Port of Rotterdam survey and monitoring non-native species conform HELCOM/OSPAR protocol. GiMaRIS report 2014_31

[8] Faasse M A Ten Hove H A and Paalvast P 2020 The Pacific serpulid tube worm *Hydroides ezoensis* Okuda, 1934 reaches the ports of Rotterdam and Vlissingen (Flushing), the Netherlands (North Sea) *Cahiers de Biologie Marine* **61** 415-421. 10.21411/CBM.A.F2973099

[9] Paalvast P, van Wesenbeeck B K, van der Velde G, and de Vries M B 2012. Pole and pontoon hulas: an effective way of ecological engineering to increase productivity and biodiversity in the hard-substrate environment of the port of Rotterdam. *Ecological engineering* **44** 199-209. https://doi.org/10.1016/j.ecoleng.2012.04.002

[10] Anastasopoulou A, Mytilineou C, Smith C J, and Papadopoulou K N 2013 Plastic debris ingested by deep-water fish of the Ionian Sea (Eastern Mediterranean). *Deep Sea Research Part I: Oceanographic Research Papers* **74** 11-13 https://doi.org/10.1016/j.dsr.2012.12.008

[11] Limongi P, Lacerot G and Segura A 2019 Plastic fibres in the gastrointestinal tract content of two South Atlantic coastal fish species with different trophic habits (Urophycis brasiliensis, Paralonchurus brasiliensis) in Punta del Diablo-Uruguay *Pan-American Journal of Aquatic Sciences* **14** 71-76 https://panamjas.org/pdf_artigos/PANAMJAS_14(1)_71-76.pdf

[12] Rochman C M, Hoh E, Kurobe T and Teh S J 2013 Ingested plastic transfers hazardous chemicals to fish and induces hepatic stress. *Scientific reports* **3** 1-7 https://doi.org/10.1038/srep03263

IOP Conf. Series: Materials Science and Engineering 1245 (2022) 012006 doi:10.1088/1757-899X/1245/1/012006

Effects of material composition and face block exposure on the long-term (2014-2018) colonisation of an intertidal RECIF Artificial Reef in the Bay of Seine

J C Dauvin[1], M Deloor[1,2], J P Pezy[1], A Raoux[1], P Claquin[2] and A Foveau[3]

[1] Normandie Univ., UNICAEN, CNRS UMR 6143 M2C, Laboratoire Morphodynamique Continentale et Côtière, UMR CNRS 6143 M2C, 24 rue des Tilleuls, F-14000 Caen, France;
[2] Normandie Univ., UNICAEN, UMR CNRS, BOREA, Biologie des ORganismes et Ecosystèmes Aquatiques (BOREA, UMR 8067), Sorbonne Université, Muséum National d'Histoire Naturelle, CNRS, Université Pierre et Marie Curie, Université de Caen Normandie, IRD 207, Université des Antilles, Centre de Recherches en Environnement Côtier (CREC)-Station Marine BP49, 54, rue du Docteur Charcot, F - 14530 Luc-sur-Mer, France
[3] Ifremer, Laboratoire Environnement Ressources Bretagne Nord, 38 Rue du Port Blanc, F-35800 Dinard, France.

Abstract. An experimental artificial reef (AR) with the incorporation of crushed seashells of the queen scallop *Aequipecten opercularis* was initiated on the intertidal zone of the Bay of Seine in the framework of the INTERREG IVa RECIF project for a everyone year monitoring observations from March 2014 to April 2018. Three block types were operated: eco-friendly material with 20% or 40% of crushed queen scallop shells, and ordinary concrete made from natural aggregates. On the seven blocks analysed at the end of the experiment in 2018, 74 taxa including 32 sessile and 42 motile fauna have been accounted; no difference of taxonomic richness was observed between material and immersion time. The sessile fauna was identified for the six external faces of two blocks aged of four years: horizontal above face (FAB), horizontal below Face (FBE), face oriented offshore (FOF), face oriented inshore (FIN), vertical face oriented west (FEW) and vertical face oriented east (FEA). FBE showed higher taxonomic richness and abundance than the other block faces. FAB exposed to the light showed the lowest abundances. Our observations showed the efficiency of such AR on the intertidal zone accessible at each spring tide, which was easier to survey than in the subtidal zone, where divers must be mobilised to sample blocks.

1. Introduction

Artificial Reefs (ARs) have been used for decades around the world to create, protect or restore a rich and diverse ecosystem [1-5]. Furthermore, recent deployment on ARs were focused on community structure or composition of ARs, showing a shift from improving fisheries as a resource to marine ecosystem rehabilitation and marine structure increasing the biodiversity [4, 6-8].

One of the scientific goals regarding ARs are to identify the species colonizing a reef, i.e. the biodiversity, the temporal succession of species after its deployment and the growth of existing species [1, 9-10]. Many motile taxa were associated with the sessile fouling organism, which increases the attractiveness of such ARs for predators [11]. Nevertheless, most of the monitoring's of ARs considered only the species composition and succession of main groups of macroalgae or macrofauna, mainly sessile organisms, and during a short period frequently one year [7, 12-13]. Only few ARs were surveyed during a long-term period (> 3 years) [7, 14-15].

Nowadays, the use of shells derived from oyster culture or king scallop fisheries, which were considered as waste, was also a request of sustainable development in material construction of ARs [16]. In this perspective, the 2013-2015 RECIF project proposed the incorporation of crushed seashells of the queen scallop *Aequipecten opercularis* (Linnaeus, 1758) into the substrate of concrete blocks as an innovative building materials development for ARs [16]. In this project, an experimental structure was initiated in March 2014 on the intertidal zone of the Bay of Seine (eastern part of the English Channel) [9-10, 17]. The short-term colonisation during the first year of monitoring until February 2015 was described in [9]. At the end of the one-year experiment (2014-2015), it was chosen to extend the monitoring every year until April 2018, for a four-year temporal monitoring of the colonisation and succession of such innovative intertidal ARs in the English Channel [18].
The objectives of this paper are: 1) to provide the inventory of sessile and motile invertebrates found in April 2018 four years after the beginning of the monitoring; and 2) to identify the effect of the material composition of the motile and sessile fauna on block colonization efficiency, and 3) to identify the colonisation pattern of the sessile fauna on the six external faces of the blocks according to their position and their location on the oyster tables.

2. Materials and Methods

2.1. Experiment site and design of blocks

At the beginning of the experiment (19-20 March 2014), 75 blocks (20 x 20 x 40 cm) were deployed on eight oyster culture tables on the intertidal zone of Luc-sur-Mer (49°19′15″N-0°20′55″W) [9-10, 17] (Figure 1). At high tide, the water depth is 6.5 m and the concrete blocks were emerged about 44% of the time [9].

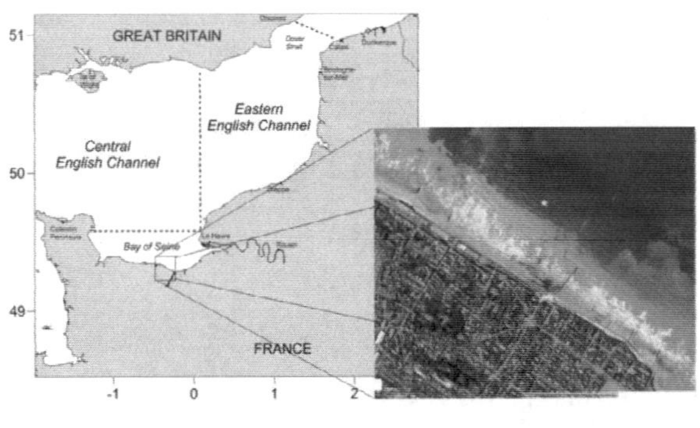

Figure 1. Location of the site at Luc-sur-Mer (Calvados coast, southern part of the Bay of Seine) [9].

Three types of block have been studied: 60 blocks of eco-friendly material for each with 40% (30 blocks) and 20% (30 blocks) of crushed queen scallop *Aequipecten opercularis* (L., 1758) shells and, and 15 blocks of ordinary concrete made from natural aggregates [15]. Then in a view to a yearly temporal monitoring, the blocks were collected once a year until April 2018. On the seven blocks collected in 2018 (Figure 2), one corresponded to ordinary concrete, two with 20% and four with 40 % of crushed queen scallop shells. Two blocks had resided at sea during four years, one during three years, two during two years and the two last blocks only during one year.

Figure 2. Blocks on oyster tables at the end of the experiment in April 2018.

2.2. Laboratory analyses of the blocks

The sessile and motile fauna were examined in the laboratory with two distinguished protocols. The sessile fauna was observed under a binocular microscope. Sub-sampling was carried out over scraping unit areas delimited by a mask for each of the six external faces: horizontal above face (FAB), horizontal below Face (FBE), face oriented offshore (FOF), face oriented inshore (FIN), vertical face oriented west (FWE) and vertical face oriented east (FEA) of each block (Figure 3). This process was used for the two four years blocks sampled in April 2018, i.e. natural and 20% of crushed queen scallop shells.

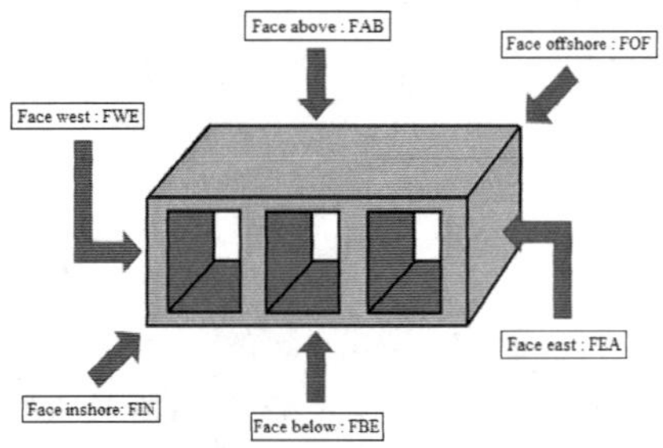

Figure 3. Designations of the six faces of the experimental blocks.

On horizontal above face (FAB) and below (FBE): 8 x 0.0025 m^2 for a total of 0.02 m^2 corresponding to 1/4 of the total surface of each horizontal face had been analysed. For the vertical faces oriented respectively to the east (FEA) and the west (FEW): 4 x 0.0025 m^2 for a total surface of 0.01 m^2, corresponding to 1/4 of the total surface of each vertical face were analysed. On the two vertical faces (not full) oriented offshore (FOF) and inshore (FIN): 2 x 0.003 m^2 and 1 x 0.004 cm^2 corresponding to 1/3 of the total surface of each vertical face had been analysed.

Moreover, each face was photographed in its entirety as well as each sub-sample. Countable organisms (barnacles, mussels, some solitary ascidians) had been considered for quantitative analyses. For each face, an inventory of all present taxa was carried out, as well as counted where it was possible. All organisms were stored in ethanol 96°.

After having placed the blocks in bowls in anoxic conditions at least 24h, the seawater used to keep the fauna alive during the laboratory observations was then filtered on a 0.5-mm mesh sieve to collect the motile fauna. The retained material was fixed with 96% alcohol, and then it was identified to the more precise level of taxonomy and count as motile fauna living associated with the blocks. For polychaetes and nemerteans, only the head of the individuals were counted.

All the abundances of sessile and motile organisms had been normalized to 0.6 m^2.

2.3. Statistical analyses
To test the differences of colonisation between block faces, taxonomic richness (TR) and total abundances (0.6 m²) for the sessile and motile, and all taxa combined were compared by ANOVAs. For the sessile fauna, the comparison was based on block faces (FAB, FBE, FOF, FEW, FIN, and FEA). For motile fauna, only the total fauna (taxonomic richness and abundances) between blocks have been tested. The null hypotheses H_0 stated that there was no effect of a factor or no effect of an interaction between factors on any of the variables. The alternative assumptions H_1 stated that there were effects of the date, or the faces. A post-hoc test (Tukey) was performed if the H_0 null hypothesis was rejected in favour of the H_1 hypothesis. The normality and homogeneity of the variances in the distribution of the data or residuals were tested with a Shapiro test and Bartlett tests, respectively. When one of these conditions was not meet, a non-parametric Scheirer-Ray-Hare test was applied instead of the ANOVA, followed by a post-hoc Dunn test.

3. Results

3.1. *Composition of the fauna in 2018*
Within the seven blocks analysed in 2018, the sessile fauna accounted 32 taxa while the motile fauna accounted 42 taxa for a total of 74 taxa. Among the sessile fauna, the most diversified groups were the polychaetes, ascidians, demosponges, and cnidarians mainly belonging to the group of hydrozoans (Table 1). However, in terms of abundance, the most numerous countable sessile species were the crustacean barnacle *Balanus crenatus*, *Perforatus perforatus*, the bivalve *Mytilus edulis*, the ascidia *Perophora japonica, Corella eumyota* and *Molgula* spp. as well as the polychaetes *Spirobranchus* (Table 1). Among the non-countable sessile species, the hydrozoans *Kirchenpaueria pinnata*, *Dynamena pumila* and *Obelia longissima* as well as the sponges *Halichondria panicea* and *Hymeniacidon* sp. were presented in all the faces of the blocks and covered in some cases the totality of a face. Among the motile fauna, polychaetes remained the most diversified group, followed by malacostraca (mainly amphipods, decapods and isopods). There were few differences between the TR of the two kinds of blocks (Table 1). The most abundant motile taxa were the Nematoda, the polychaete Nereidae and Phyllodocidae *Eulalia*, and the tanaid *Zeuxo holdichi* (Table 1).
The total number of taxa per blocks varied from 31 for the four year block with natural aggregates and 44 for a one year block with 40% of crushed queen scallop shells (Table 1). The number of sessile taxa was on the same order of magnitude on the seven blocks: i.e. from 16 to 19 without significant differences between years (p=0.343), while the number of motile taxa varied strongly from 16 to 29 but without significant difference (p=0.352). It was notable that the number of taxa of the four years blocks was among the lowest values (Table 1). The number of taxa on the seven blocks analysed in 2018 showed no significant differences between blocks (p=0.819 for sessile taxa and p=0.304 for motile taxa)
In summary, they were no significant differences on temporal and nature of aggregates patterns. The colonisation was fast in one year. Then low changes on sessile taxa number were observed while those of motile taxa varied highly from one block to another without link to the immersion time (Table 1).

3.2. Colonisation pattern according to the faces of the blocks
The figure 4A showed the number of sessile taxa found on each face of both 4 year blocks. The number of taxa varied from a minimum of two to a maximum of 12; but there were no significant differences between faces (p=0.249).

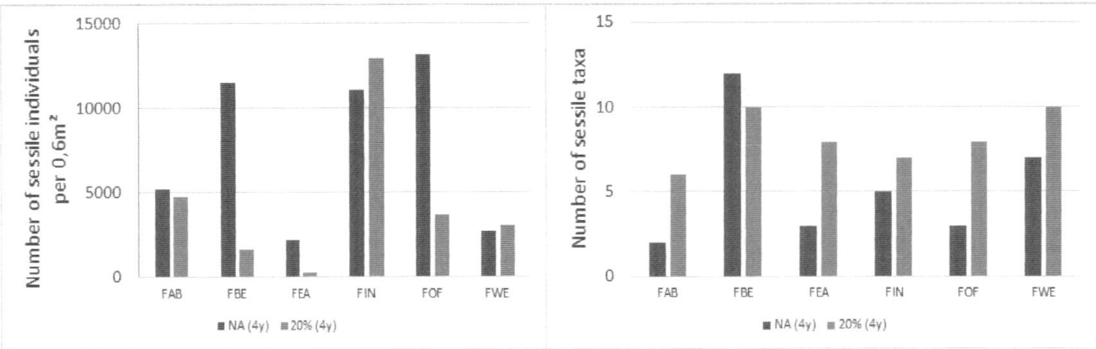

Figure 4. Number of individuals (left) and number of taxa (right) of sessile taxa accounted in the six faces of the two 4 year blocks (NA: Natural aggregate and 20%: 20% of crushed queen scallops). FIN: face oriented inshore; FOF: face oriented offshore, FBE: horizontal below Face; FAB: horizontal above face; FWE vertical face oriented west (FEW) and FEA: vertical face oriented east.

The figure 4B showed the abundance per 0.6 m^2 of sessile and motile fauna in the six faces of the two 4 year blocks. The western and eastern faces showed low abundances while the face oriented inshore and offshore showed high abundances in the same order of magnitude than of the below face. The horizontal above face showed intermediate numbers of individuals between maximal and minimal values; but there were no significant differences between faces ($p=0.235$).
The figure 5 illustrated the difference of colonisations of two main faces the above and the below faces for both 4-year blocks with the presence of algae on the above faces while the below faces were mostly colonised by sessile taxa.

4. Discussion

Analyses of the short-term colonization (2014-2015) showed temporal difference in species abundance with a succession of macroalgae and macrofauna colonizers along the first year of the experiment, but there were no difference in colonization according to the block composition [10]. Moreover, during the first year of the experiment, the main colonizing organisms were barnacles, hydrozoans, tunicates, mussels and sessile annelids with a biological succession over time: a first assemblage mainly dominated by barnacles then a richer assemblage with the dominance of ascidians [10]. Our monitoring confirmed the permanence and dominance of the barnacles mainly the species *Balanus crenatus* and *Perforatus perforatus,* the mussel *Mytilus edulis,* the Polychaeta *Spirobranchus*, and the ascidian - among them *Corella euryota* and *Perophora japonica* were Non-Indigenous Species in Normandy. Among the motile fauna, the blocks were colonized by small species such as Nematoda, the polychaeta Nereidae, Cirratulidae, Syllidae, *Eulalia* spp., and crustacean such as the isopod *Dynamene bidentata*, *Gnathia* spp., the amphipod *Monocorophium* and the tanaid *Zeuxo holdichi*. This tanaid was very abundant during the first year of the experiment with several thousands of individuals collected and until 2,500 ind.m^{-2} in 2015 [10]. The species could be classified as pioneer and invasive species, with dramatically decreased along the time on artificial structures. The reason for the high abundances in the first stage of block colonisation at this location remained enigmatic.

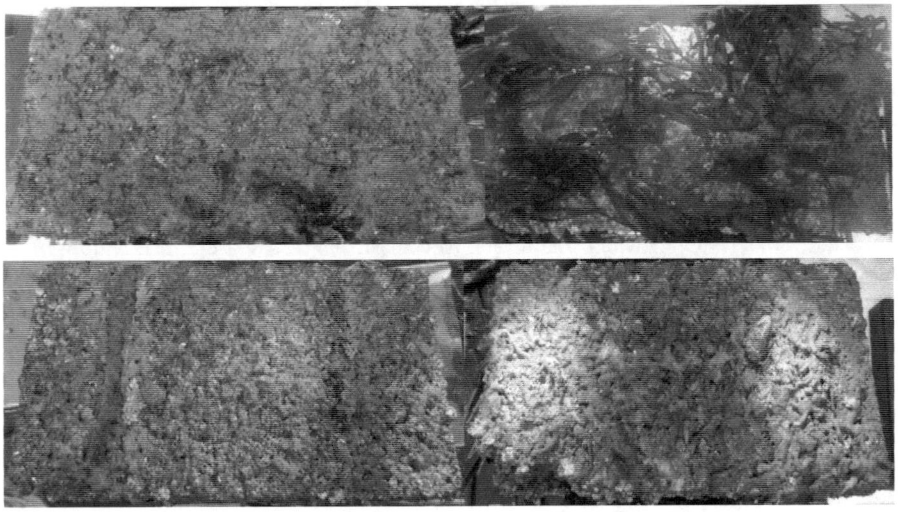

Figure 5. Photos of the four year block. On left 20% crushed queen scallop and on right Natural aggregates. Above FAB: horizontal above face and below FBE: horizontal below Face.

During the first year of immersion, the main environmental factors controlling settlement, colonization and competition between species were : speed of tidal currents, wind force and direction, sunshine hours and rainfall during emergence with species settled preferentially on certain faces according to environmental parameters (light, hydrodynamics) and larval behaviour [10, 17]. A significant taxonomic colonisation was observed mainly on the above and below faces of the blocks. It was clear that the light, the desiccation of the exposed above face of the block and the protection of the below faces on the block played an important role in the colonisation of the blocks, which is amplified in our intertidal experiment with the alternation of immersion and emersion phase.
Our four years experiment showed that the colonisation of such blocks deployed on oyster table on the intertidal zone of the Bay of Seine was very fast; in one year the taxonomic richness underpasses 40 and remained stable along the time. Butler and Connolly (1995) [19] surveyed long term dynamics of sessile fauna, in the Spencer Gulf, South Australia, during six years after initial immersion of piles and showed that the taxonomic composition were stable after three years of immersion. For a long-term (1981-2001) experiment, Nicoletti et al. (2007) [14] identified five phases of colonisation for ARs deployed in the mid Tyrrhenian Sea, Italy, from pioneer species during the first months of immersion to Bryozoans bioconstruction dominance at the end of the experiment, and with phase of dominance, regression and absence of the mussel *Mytilus galloprovincialis*. Recently, Taormina et al. (2020) [20] had monitored epibenthic colonization of artificial structure in the subtidal North-Brittany (France) coast using image-based underwater monitorings and showed a mature stage in four years.
Finally, our experiment showed the efficiency of such AR on the intertidal zone accessible at each spring tide, which was easier to survey than in the subtidal zone, where divers must be mobilised to sample blocks, and can be easily deployed to test for example different aggregates used in the ARs construction and design.

Acknowledgements
This work was supported by the collaborative projects RECIF and MARINEFF selected under the European Cross-border Programme INTERREG FMA cooperation between France and the United Kingdom in adjacent regions around the English Channel, and co-funded by the ERDF. It forms part of the REGENI (Réalisation d'un Guide des Espèces Non-Indigènes) and ENBIMANOR (ENrichissement de la BIodiversité MARine en NORmandie) projects financed by the AESN (Agence de l'Eau Seine Normandie). The authors wish to thank the co-financiers and all project partners for their support.

IOP Conf. Series: Materials Science and Engineering 1245 (2022) 012006 doi:10.1088/1757-899X/1245/1/012006

References

[1] Jensen A, Collins KJ, and Lockwood APM, editors. 2000, Artificial reefs in European seas. Springer Science and Business Media, B. V. Dordrecht. Boston:*Kluwer Academic Publishers*.

[2] Fabi G, Spagnolo A, Bellan-Santini D, Charbonnel E, Çiçek BA, Goutayer García JJ, Jensen AC, Kallianiotis A and Neves dos Santos M. 2011. Overview on artificial reefs in Europe. *Braz. J. Oceanogr.* **59**, 155-166.

[3] Tessier A, Francour P, Charbonnel E, Dalias N, Bodilis P, Seaman W, and Lenfant P 2015, Assessment of French artificial reefs: due to limitations of research, trends may be misleading. *Hydrobiologia* **753**, 1-29.

[4] Lee MO, Otakeb S and Kima JK 2018, Transition of artificial reefs (ARs) research and its prospects *Ocean Coast. Manag.* **154**, 55-65.

[5] Vivier B, Dauvin JC, Navon M, Rusig AM, Mussio I, Orvain F, and Claquin P 2021, Marine artificial reefs in the world, a literature analysis of their designs, objectives and effectiveness. *Global Ecol. Distr.* **27**, e01538.

[6] Pickering H, Whitmarsh D, and Jensen A 1999, Artificial reefs as a tool to aid rehabilitation of coastal ecosystems: Investigating the potential. *Mar. Poll. Bull.* **37**, 505-514.

[7] Perkol-Finkel S, and Benayahu Y 2005, Recruitment of benthic organisms onto a planned artificial reef: shifts in community structure one decade post-deployment. *Mar. Env. Res.* **59**, 79-99.

[8] Perkol-Finkel S, Shashar N, and Benayahu Y 2006, Can artificial reefs mimic natural reef communities? The roles of structural features and age. *Mar. Env. Res.* **61**, 121-135.

[9] Foveau A, Dauvin JC, Rusig AM, Mussio I, and Claquin P 2015, Colonisation à court terme par le benthos sur un éco-récif artificiel. In: Boutouil M, Leboulanger S, editors. Proceedings of the RECIF Congress on artificial reefs: from materials to ecosystems. ESITC Caen, France, ISBN 97862695517664606\5, p 119-126.

[10] Dauvin JC and Foveau A 2019, One-year colonisation by zoobenthic species on an ecofriendly artificial reef in the English Channel intertidal zone. In: Komatsu T, Ceccaldi HJ, Yoshida J, Prouzet P, and Henocque Y (eds) *Oceanography Challenges to Future Earth, Springer*. p 285-294.

[11] Moreau S, Péroni C, Pitt KA, Connolly RM, Lee SY, and Méziane T 2008, Opportunistic predation by small fish on epibiota of jetty pilings in urban waterways. *J. Fish Biol.* **72**, 205-217.

[12] Brown CJ. 2005. Epifaunal colonization of the Loch Linnhe artificial reef: influence of substratum on epifaunal assemblage structure. *Biofouling* **21**: 73-85.

[13] Boaventura D, Moura A, Leitao F, Carvalho S, Curdia J, Pereira P, Cancela da Fonseca L, Neves dos Santos M and Costa Monteiro C 2006, Macrobenthic colonisation of artificial reefs on the southern coast of Portugal (Ancao, Algarve). *Hydrobiologia* **555**, 335-343.

[14] Nicoletti L, Marzialetti S, Paganelli D, and Ardizzone GD 2007, Long-term changes in a benthic assemblage associated with artificial reefs. *Hydrobiologia* **580**, 233-240.

[15] Carvalho S, Moura A, Cúrdia J, Cancela da Fonseca L, and Santos MN 2013, How complementary are epibenthic assemblages in artificial and nearby natural rocky reefs? *Mar. Environ. Res.* **92**, 170-177.

[16] Cuadrado H, Boutouil M, Boudart B, Claquin P, and Leroy F 2016, Colonisation et détérioration des bétons incorporant des coquilles pour récifs artificiels. *Mat. Tech.* **104**, 1-11.

[17] Claquin P, Leroy F, Rusig AM, Mussio I, Feunteun E, Foveau A, Dauvin JC, Gallon R, Le Brun JL, Lestarquit M, Orvain F, Martinez A, Desoche E, Napoléon C, Roussel D, and Boutouil M. 2015, Récif artificiel : mise en place d'un suivi de la colonisation à plusieurs échelles. In: Boutouil M, Leboulanger S, editors. Proceedings of the RECIF Congress on artificial reefs: from materials to ecosystems. ESITC Caen, France, ISBN 978626955176646065, p 119-126.

[18] Dauvin JC, Deloor M., Pezy JP, Raoux A, Claquin P, and Foveau A submitted. Four years-temporal survey of an intertidal Artificial Reef in the English Channel. *Biofouling*

[19] Butler AJ and Connolly RM 1996, Development and long term dynamics of a fouling assemblage of sessile marine invertebrates. *Biofouling* **9**, 187-209.

[20] Taormina B., Percheron A., Marzloff M. P., Caisey X., Quillien N., Lejart M., Desroy N., Dugornay O., Tancray A., and Carlier A 2020, Succession in epibenthic communities on artificial reefs associated with marine renewable energy facilities within a tide-swept environment. *ICES J. Mar. Sci.*

Table 1. List of the taxa recorded in the seven blocks studied in April 2018. NA: natural aggregates, 20% and 40% aggregates with 20% or 40% of crushed queen scallop shells. Y: number of immersion year. Black case: presence of the taxon and number corresponding to the abundance per 0.6 m². * Non-Indigeneous Species.

Motile fauna		40% (1y)	20% (1y)	40% (2y)	40% (2y)	40% (3y)	NA (4y)	20% (4y)
Ammothella longipes	Hodge, 1864	■	■		■	■	1	1
Amphipholis squamata	Delle Chiaje, 1828						-	2
Arenicolides ecaudata	Johnston, 1835			■			-	1
Cancer pagurus	Linnaeus, 1758	■		■		■	-	
Carcinus maenas	Linnaeus, 1758	■	■	■	■	■	-	1
Chironomidae					■		-	-
Cirratulidae	Ryckholt, 1851	■				■	6	43
*Monocorophium**	Latreille, 1806			■			10	-
Dynamene bidentata	Adams, 1800	■	■	■	■	■	2	10
Eulalia	Savigny, 1822	■	■	■	■	■	36	-
Eumida sanguinea	Örsted, 1843	■		■		■	1	-
Eunicida					■		-	-
Fecampia erythrocephala	Giard, 1886	■		■		■	-	-
Gnathia	Leach, 1814	■	■	■	■	■	27	5
Golfingia vulgaris	de Blainville, 1827						-	2
Harmothoe	Kinberg, 1856	■				■	-	-
Lekanesphaera monodi	Arcangeli, 1934	■				■	-	-
Lepidonotus squamatus	Linnaeus, 1758	■				■	-	6
Lineus	Sowerby, 1806	■		■		■	14	5
Marphysa sanguinea	Montagu, 1813	■				■	-	-
Microdeutopus anomalus	Rathke, 1843	■				■	-	-
Nematoda		■	■	■	■	■	61	287
Nemertea		■	■	■	■	■	11	5
Nereididae	Blainville, 1818	■	■	■	■	■	112	5
Nototropis swammerdamei	Milne Edwards, 1830				■	■	-	-
Nucella lapillus	Linnaeus, 1758	■				■	-	-
Ostracoda	Latreille, 1802	■	■	■		■	-	3
Peringia ulvae	Pennant, 1777						-	1
Pholoe inornata	Johnston, 1839	■		■		■	11	12
Phoxichilidium femoratum	Rathke, 1799						-	-
Phyllodoce mucosa	Örsted, 1843		■	■			-	-
Phyllodocidae	Örsted, 1843		■		■		-	-
Pilumnus hirtellus	Linnaeus, 1761	■	■	■	■	■	-	-
Pinnotheres pisum	Linnaeus, 1767				■		-	-
Polititapes rhomboides	Pennant, 1777	■		■	■	■	-	-
Polynoidae	Kinberg, 1856	■	■		■	■	-	-
Porcellana platycheles	Pennant, 1777	■	■	■	■	■	-	-
Schistomeringos	Jumars, 1974			■			-	-
Syllidae	Grube, 1850	■	■	■		■	10	16
Tryphosa nana	Krøyer, 1846	■				■	-	-
Venerupis corrugata	Gmelin, 1791		■				1	1
Zeuxo holdichi	Bamber, 1990	■		■		■	54	-
Total of motile fauna		**28**	**16**	**32**	**17**	**29**	**15**	**18**
Sessile fauna								
Balanus crenatus	Bruguière, 1789	■	■	■	■	■	4505	2064
Bispira polyomma	Giangrande & Faasse, 2012	■	■	■		■	6	2
Botrylloides	Milne Edwards, 1841						-	■
Botryllus schlosseri	Pallas, 1766						-	■
Branchiomma bombyx	Dalyell, 1853	■	■	■	■	■	9	21
Ciona	Fleming, 1822	■	■	■	■	■	-	-
*Corella eumyota**	Traustedt, 1882	■	■	■	■	■	4	80
Diadumene cincta	Stephenson, 1925						9	

IOP Conf. Series: Materials Science and Engineering 1245 (2022) 012006 doi:10.1088/1757-899X/1245/1/012006

Diphasia fallax	Johnston, 1847						■	■
Dipolydora	Verrill, 1881						-	1
Dynamena pumila	Linnaeus, 1758				■	■	-	-
Flustrellidra hispida	Fabricius, 1780						-	4
Halichondria	Fleming, 1828		■	■	■	■	■	■
Hymeniacidon	Bowerbank, 1858	■	■	■	■		■	■
Jasmineira elegans	Saint-Joseph, 1894	■		■		■	-	-
Kirchenpaueria pinnata	Linnaeus, 1758		■	■	■	■	■	-
Lanice conchilega	Pallas, 1766	■				■	2	-
Molgula	Forbes, 1848	■	■	■	■	■	40	72
Musculus discors	Linnaeus, 1767				■		-	-
Mytilus edulis	Linnaeus, 1758	■	■	■	■	■	29	106
Obelia longissima	Pallas, 1766						■	■
Perforatus perforatus	Bruguière, 1789	■	■	■	■	■	71	52
*Perophora japonica**	Oka, 1927	■					64	84
Pista mediterranea	Gaillande, 1970			■			-	-
Polyclinum	Savigny, 1816						-	-
Psamathe fusca	Johnston, 1836			■			-	-
Sabellidae	Latreille, 1825	■	■	■	■	■	-	-
Serpulidae	Rafinesque, 1815	■	■			■	-	-
Spirobranchus	Blainville, 1818	■	■	■	■	■	56	31
Suberites	Nardo, 1833		■	■	■	■	-	-
Terebellidae	Johnston, 1846	■	■	■	■	■	-	1
Urticina	Ehrenberg, 1834						-	2
Total of sessile taxa		**16**	**16**	**18**	**16**	**18**	**17**	**19**
Total of taxa		**44**	**32**	**40**	**33**	**47**	**31**	**37**

Enhancing marine coastal biodiversity in Normandy: towards a new species signalling network using Artificial Structures

J P Pezy[1], A Raoux[1], R Boisserie[1] and J C Dauvin[1]

[1]*Normandie Univ., UNICAEN, CNRS UMR 6143 M2C, Laboratoire Morphodynamique Continentale et Côtière, UMR CNRS 6143 M2C, 24 rue des Tilleuls, F-14000 Caen, France*

Abstract. Non-Indigenous Species (NIS) are a threat to native biodiversity and ecosystem services. The European seas are known to be the recipient of several hundreds of NIS due two main origins: shipping and aquaculture. They are the focus of Descriptor 2 of good environmental status in the Marine Strategy Framework Directive. 153 NNS had been recorded along the Normandy coast in 2018. Marinas are priority sites for monitoring, being 'hotspots' for the NIS occurrence. The aim of the ENBIMANOR project was to study and compare the composition and the hard-bottom macrofauna including the NIS prevalence in marinas along the Normandy coast (from Granville to Dieppe-Le Tréport) focusing primarily on mobile and benthic invertebrates. A total of 19 marinas were surveyed during two years (2018-2020). At each harbour, nine settlements panels (acting as artificial reef) were immerged at a depth of 1.50 m. Every quarter, at each site, a settlement panel was replaced and the species colonising the panel were identified. The analysis revealed higher taxonomic richness in marinas open to the sea or with high salinity than in closed marinas or with low salinity. Results highlighted that each marina has its own species composition. This survey identified more than one hundred taxa, including known NIS species such as the amphipod *Monocorophium acherusicum* and *M. sextonae,* the ascidians *Perophora japonica* and *Stylea clava,* and detect four new NIS for the Normandy waters (*Aoroides longimerus, A. curvipes, Paranthura japonica* and *Ianiropsis serricaudis).*Our study shows the need to continue the census of NIS and to evaluate their impacts on the artificial hard-bottom structures in marine coastal ecosystem on two major zones: the intertidal zone and the marina/harbours waters.

1. Introduction

Intentional or unintentional species introductions by human activities to regions outside their native ranges have been on-going for thousands of years [1]. However, these ecosystem invasions by Non-Indigenous Species (NIS) are considered nowadays as one of the greatest threats to biodiversity world-wide [2,3]. NIS introductions into coastal zones have been accelerated thorough the world in recent decades mainly due to increased human activities such as shipping, aquaculture, maritime traffic and tourism [4]. This increase of NIS introduction occurring in the context of global climate change leads to modifications in the ecosystem structure and function with potential repercussions on their ability to provide goods and services to humans [2]. In fact, NIS presents a wide range of threats to native ecosystems, and can be responsible for the decline of native species by competition and predation, and thus can lead to a loss of biodiversity [5]. Moreover, NIS can also cause direct economic impacts on human activities structure and aquaculture. Invasive species can also cause disappearance of the original communities. A famous example is the invasion of the ctenophore *Mnemiopsis leidyi* in the Black Sea, which led to a collapse of pelagic fisheries. Finally, it is worth to note that in the marine environment, these introductions are most of the time irreversible. Long term dataset analyses on NIS can provide insights into past and future trends for conservation purposes [6]. However, such monitoring surveys are relatively rare [6]. In this context, the ENBIMANOR project (ENhancing MArine, coastal BIodiversity in NORmandy: towards a new species signalling network) aims to understand the evolution in biodiversity of the Normandy coastline and marinas. This on-going project analyses the fixed hard-substrate fauna present in marinas and will test the hypothesis that the ports of Normandy with high numbers of cross-Channel and international traffic (Cherbourg, Ouistreham, Le Havre et Dieppe) are the main pathway of introduction of non-indigenous (but rarely invasive) species.

2. Materials and Methods

2.1 Experimental designs

We selected 19 marinas with floating pontoons located along the 640 km long coastline of Normandy, from Granville to Le Tréport (Figure 1). These marinas were representative of a range of characteristics (open to the sea or closed, fully marine or under freshwater (estuarine) influence). The species settlement dynamics were investigated in these 19 marinas. In each marina, nine panels (25 cm x 25 cm; 0.065 m²; Figure 1) were immersed at 1.5 m depth under a pontoon. Every quarter, in each marina, plates were removed and analysed after 3, 6, 9, 12, 15, 18 and 24 months, in order to view the entire colonisation sequence. The retrieved specimens of each plate were preserved in alcohol solution and taken to the laboratory for identification. Then organisms were sorted, identified and counted using a binocular microscope and stored in alcohol. In addition, each month in each marina, a CTD probe was used to monitor standard physico-chemical parameters (temperature, salinity, turbidity and concentration of dissolved oxygen). In this study, we only present the main results corresponding to the monitoring period of 15 months. The physico-chemical parameters were presented for 18 marinas during the 15 months (Saint Vaast la Hougue was equipped only after 12 months). For the biological analysis, the site of Le Tréport was excluded (absence of fauna), thus, the analyses were made on 17 marinas.

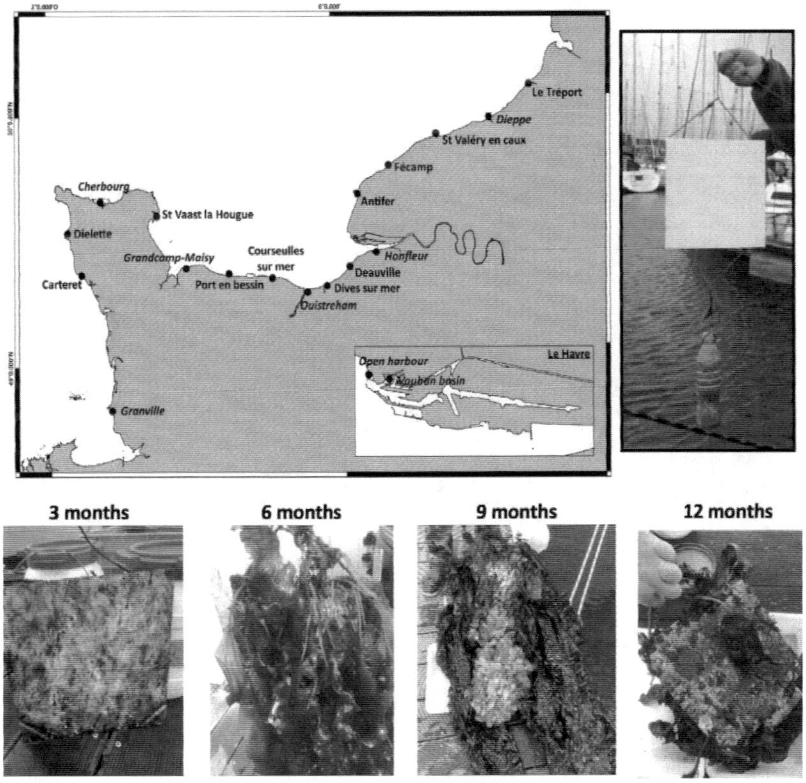

Figure 1. Location of the 19 marinas selected and photography's from Cherbourg from 3 to 12 months of plate colonisation.

IOP Conf. Series: Materials Science and Engineering 1245 (2022) 012007 doi:10.1088/1757-899X/1245/1/012007

2.2 Data analysis

First, the typology of the 18 marinas was investigated using the multivariate analysis, a Hierarchical Ascendant Classification (HAC) was created by means of group-average linking using the Bray-Curtis similarity measure.

Then, species names were checked with World Register of Marine Species (WORMS, http://www.marinespecies.org) on June 2019. The community structure analysis was based on the taxonomic richness, the Shannon-Wiener diversity index H′ using $\log_2$ base and the Pielou evenness index J′. Both indices were calculated for plates in each marina, for a total of 17 marina.

The spatial structure of the benthic community of the 17 marinas was investigated using the multivariate analysis method of Clarke and Warwick (1994) with the PRIMER-6 software package (Plymouth Routines In Multivariate Ecological Research). Data analyses were performed by using non-metric multidimensional scaling ordination (nMDS), and a Hierarchical Ascendant Classification (HAC) was created by means of group-average linking using the Bray-Curtis similarity measure. $\mathrm{Log}_{10}(X+1)$-transformed data were used to down-weight the effect of the highly abundant species

3. Results

3.1 *Typology of the marinas*

Physico-chemical parameters (temperature, salinity, turbidity and concentration of dissolved oxygen) were similar in 75 % of the 18 marinas (excluding Saint-Vast-La-Hougue). A cluster analysis separates the marinas into two main groups (Figure 2). In addition, the first group "A" can be further divided into three sub-group: The sub-group "A1" corresponds to the marina located in the Seine Maritime, the sub-group "A2" comprises the marinas located along the Northwest coast of the Cotentin while the sub-group "A3" corresponds to the marinas located along the Calvados coast. The second group "B" corresponds to three following marinas: Tréport, Honfleur and Ouistreham marina.

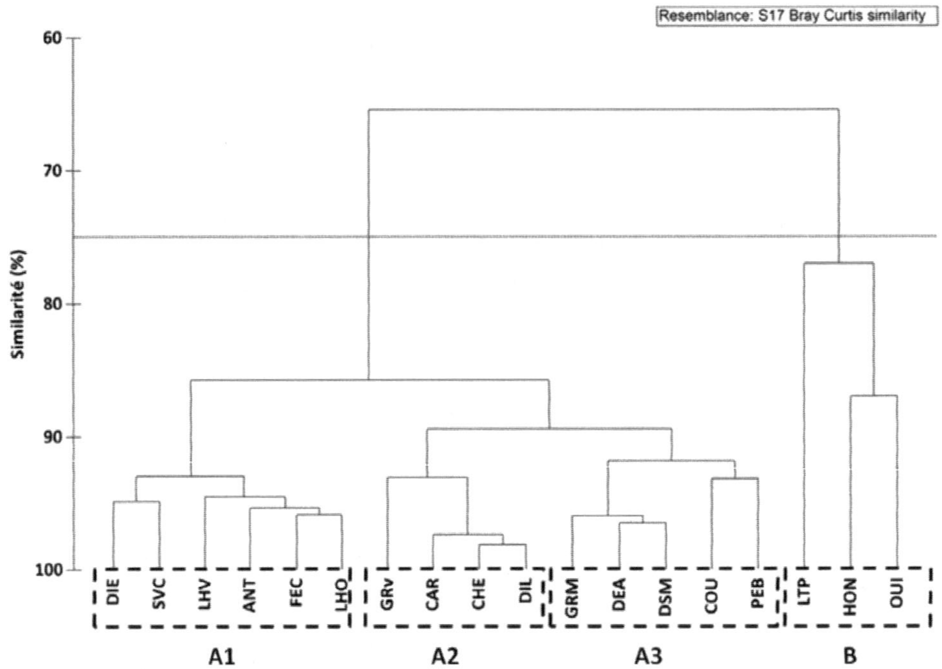

Figure 2. Cluster dendrogram showing distribution of the 18 marinas according to the physico chemical parameters.

IOP Conf. Series: Materials Science and Engineering 1245 (2022) 012007 doi:10.1088/1757-899X/1245/1/012007

Results also showed that the sub-group "A1" had the lowest average temperature with 10.88°C. The group "B" had the highest average concentration of dissolved oxygen (7.35 $mL.L^{-1}$) and the lowest average salinity (7.016) while the highest salinity values were recorded in the marinas belonging to the sub-group "A2". The highest mean turbidity was found in the sub-group "A3", with 6.88 (Table 1).

Table 1. Physico-chemical parameters (minimum, average and peaks) of the three subgroups (A1, A2, A3) and the group B

	A1	A2	A3	B
Temperature (°C)	6.7-**10.9**-17.0	7.2-**11.4**-16.0	6.7-**11.4**-16.7	6.8-**11.5**-18.1
Dissolved Oxygen (mL/L)	5.6-**6.7**-7.9	5.6-**6.2**-6.8	5.6-**6.4**-7.8	6.1-**7.3**-8.4
Turbidity (NTU)	0.4-**3.9**-12.2	0.5-**3.3**-10.3	1.4-**6.9**-24.2	0.8-**3.5**-8.0
Salinity	0.8-**23.9**-32.2	32.1-**34.1**-34.9	9.2-**28.8**-32.7	0.5-**7.0**-14.8

3.2 Biological analyses

A total of 141 taxa was recorded during the 15 months of survey of 17 marinas (excluding Saint-Vaast-La-Hougue and Le Tréport). Among these taxa, the macrofauna is dominated by arthropoda (75 taxa), following by annelida (31 taxa) and mollusca (15 taxa). Concerning the Shannon-Weaver diversity index (H'), a significant difference is observed between the early stages of settlement (three and six months) and the later stages of the study (12 and 15 months) with highest values in the later stages (Table 2). The taxonomic richness increases during the first year of immersion and a new annual cycle begins the next year. H' shows the same pattern. About abundances and Piélou index, results showed that time did not affect them. There was no difference between sites with taxonomic richness, abundance, Shannon or Piélou index. A biological succession is observed over time: a first assemblage mainly dominated by barnacle's pioneer taxa, which build new surfaces, new structures, and allowed the colonization by some other taxa. Out of the 141 taxa, 20 NIS were recorded (14.2%). It was worth to note that NISs were recorded in the 17 marinas. However out of the 20 NISs, 12 were recorded in marinas presenting a high traffic maritime (Cherbourg, Dielette, and the two Le Havre harbours).

Finally, considering only the abundance found on the panels corresponding to one year of immersion, results showed that for a similarity of 40%, we detected spatio-temporal patterns; the dendrogram separated the marinas into eight groups (Figure 3). The group « e » corresponds to marinas with high maritime traffic related to scallop fisheries. The group « f » corresponds mainly to marinas located along the Seine Maritime coasts while the group «h » corresponds to marinas located under freshwater influence. The other groups are isolated from any clustering, indicating different communities between marinas and harbours.

Table 2. Main biological characteristics of the 17 marinas. Taxonomic richness (TR), total number of species over 0.0625m² x 2. Ab (Abundance), total number of organisms over 0.0625m² x 2. H'(log2), Shannon Index. J', Pielou index. Colour-coded according to WFD (Water Framework Directive) definition of good ecological status; blue: high ecological status, green: good ecological status, yellow: moderate ecological status, orange: poor ecological status; red: bad ecological status.

Marinas	Panel	TR	Ab	H'(log2)	J'	Marinas	Panel	TR	Ab	H'(log2)	J'
ANT	P1	7	7861	0,1029	3,67E-02	GRM	P1	4	13	1,614	0,8072
	P2	14	5040	1,821	0,4783		P2	4	122	1,581	0,7906
	P3	22	4692	2,639	0,5918		P3	9	26	2,854	0,9002
	P4	24	1093	2,534	0,5526		P4	14	3143	0,522	0,1371
	P5	21	1844	2,503	0,5698		P5	15	1308	2,069	0,5296
CAR	P1	6	106	1,244	0,4812	GRV	P1	5	13	2,295	0,9886
	P2	9	587	0,7115	0,2244		P2	11	359	2,454	0,7092
	P3	17	292	2,075	0,5077		P3	14	185	3,171	0,8329
	P4	12	2288	0,8064	0,225		P4	28	321	3,645	0,7582
	P5	8	27	2,773	0,9243		P5	17	123	2,694	0,659
CHE	P1	8	26	2,637	0,879	HON	P1	2	17	0,3228	0,3228
	P2	12	558	2,484	0,6928		P2	3	20	0,9219	0,5817
	P3	19	4661	1,759	0,4141		P3	10	4083	0,7131	0,2147
	P4	21	1867	1,796	0,409		P4	16	12097	2,32	0,58
	P5	17	1859	1,45	0,3547		P5	12	1314	2,387	0,6658
COU	P1	6	41	1,973	0,7634	LHO	P1	1	2	0	****
	P2	2	196	0,1142	0,1142		P2	11	359	1,309	0,3783
	P3	10	255	1,901	0,5721		P3	26	4881	1,697	0,3611
	P4	18	1094	2,609	0,6256		P4	28	1126	2,783	0,5788
	P5	10	1096	1,581	0,476		P5	14	181	2,879	0,756
DEA	P1	3	18	0,6144	0,3876	LHV	P1	7	20	2,433	0,8667
	P2	4	13234	3,43E-03	1,72E-03		P2	10	1337	1,412	0,4251
	P3	11	1905	1,63	0,4713		P3	25	9831	1,962	0,4225
	P4	15	6173	2,075	0,5311		P4	17	1157	2,708	0,6625
	P5	19	2182	2,499	0,5883		P5	10	298	0,9508	0,2862
DIE	P1	3	100	0,4457	0,2812	OUI	P1	4	43	0,678	0,339
	P2	9	915	1,247	0,3933		P2	4	8	1,75	0,875
	P3	16	3535	2,332	0,583		P3	11	17454	0,2018	5,83E-02
	P4	23	2913	2,348	0,5191		P4	12	16624	1,263	0,3522
	P5	12	3297	0,6244	0,1742		P5	9	534	2,28	0,7193
DIL	P1	6	118	0,9989	0,3864	PEB	P1	4	40	1,794	0,8972
	P2	14	942	1,198	0,3148		P2	7	11268	7,25E-02	2,58E-02
	P3	14	1415	1,847	0,485		P3	9	357	2,384	0,752
	P4	22	6808	1,713	0,3842		P4	22	2022	2,692	0,6036
	P5	10	4046	1,516	0,4563		P5	12	416	2,412	0,6727
DSM	P1	6	9	2,503	0,9684	SVC	P1	9	325	0,6289	0,1984
	P2	9	22721	7,72E-02	2,43E-02		P2	5	306	1,303	0,5614
	P3	7	3014	0,3203	0,1141		P3	12	807	1,71	0,477
	P4	17	2458	1,667	0,4078		P4	17	431	2,929	0,7165
	P5	7	1563	1,282	0,4568		P5	20	1320	2,604	0,6024
FEC	P1	6	654	0,253	9,79E-02						
	P2	8	977	0,8901	0,2967						
	P3	25	3311	1,936	0,417						
	P4	27	3417	1,94	0,408						

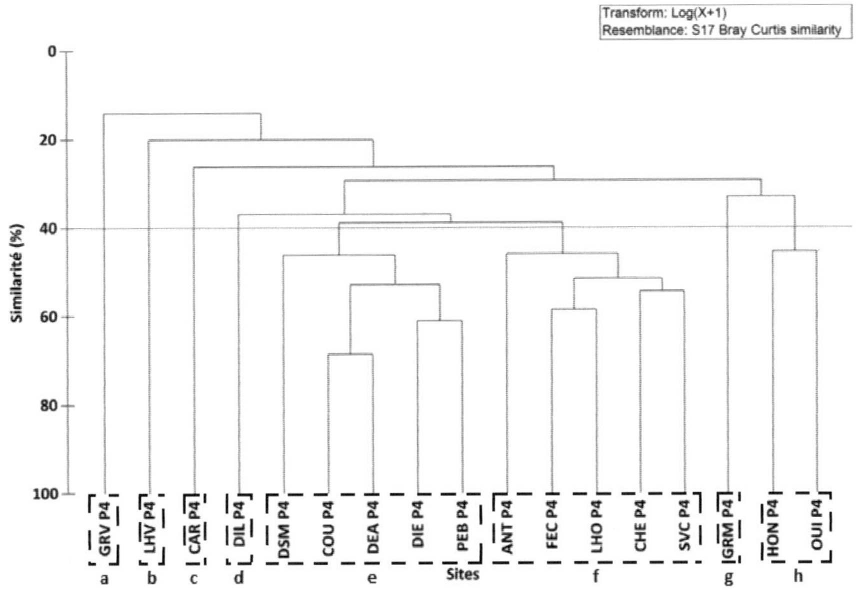

Figure 3. Cluster dendrogram showing distribution of the 17 marinas to the abundances (mean abundance after 12 months of colonisation corresponding to P4) according to the Bray-Curtis similarity after Log(x + 1) transformation of the abundance.

4. Discussion

Analyses from the physico-chemical data permitted to characterize the marinas in different groups. This typology was based on the following criteria: the number of moorings in each marina, open to the sea or closed, fully marine or under freshwater influence and the traffic maritime. According to this typology, the marinas were gathered in four groups: a group comprised of the marinas located along the west coast of the Cotentin, a group comprised of the marinas located along the Calvados coasts, a group comprised of the marinas located along the Seine maritime coasts, and a group of marina strongly influenced freshwater inflows. The analyses also highlighted higher taxonomic richness in marinas open to the sea or with high salinity than confined and / or with low salinity. It is also important to note that each port has its own unique fauna.

This survey allowed to identify more than one hundred taxa, including known NIS species such as the amphipod *Monocorophium acherusicum*, *M. sextonae,* the ascidians *Stylea clava,* the polychaete *Ficopomatus enigmaticus* and detect several new NIS for the Normandy waters (*Aoroides longimerus, A. curvipes, Paranthura japonica* and *Ianiropsis serricaudis).* For instance, high abundance of *Corophium* were observed in the 17 marinas with several thousands of individuals collected and could reach over 6,500 individuals over $0.0625m^2$ x 2. All these species originate from the north-western Pacific in Chinese and Japanese waters.

Two pathways of NIS introduction in Normandy have been proposed. The first is maritime transport; in particular Le Havre and Cherbourg. In fact, several studies pointed out that the Havre Harbor is the first place for the record of NIS in the English Channel [5]. The second pathway is aquaculture, mainly of the Japanese oyster *Crassostrea gigas* (Thunberg, 1793) and its transfer between the Atlantic and English Channel oyster centers of production [5]. Our study revealed that NISs were found in the 17 marinas and not only in the marina with high traffic maritime. This result is in line with the hypothesis of Alfonso et al. 2020 [7]; Ros et al. 2013 [8]; Chainho et al. 2015 [9]; Martinez-Laiz et al. 2019 [10] who stated that Recreational marinas are points of entry for potential colonizers and can act as stepping-stones or corridors for the spread of these species. NISs that seems to be able to disperse outside the marinas have planktonic larval stages and higher environmental tolerance, which seems to have contributed to the successful spread.

Our experiment showed the efficiency of such panels to be colonized by several marine species including NIS. In addition, the results of the present study highlighted that the colonization of such panels deployed in marina is efficient to provide an inventory of sessile and motile invertebrates living in the marinas. Finally, the results also emphasized the importance of monitoring NIS occurrence in marinas, in order to detect the arrival of new potential invaders and to prevent its dispersal to sensitive ecosystems.

5. Acknowledgements

This study forms part of the REGENI (Réalisation d'un Guide des Espèces Non-Indigènes) and ENBIMANOR (ENrichissement de la BIodiversité MARine en NORmandie) projects financed by the AESN (Agence de l'Eau Seine Normandie).

6. References

[1] Carlton JT, and Geller JB 1993, Ecological roulette, the global transport of nonindigenous marine organisms. *Science* **261**: 78-82.

[2] Vitousek P, Mooney, H Lubchenco, and Melillo J 1997, Human Domination of Earth's Ecosystems. *Science* **277**: 494-499.

[3] Mack RN, Simberloff D, Lonsdale WM., Evans H, Clout M, and Bazzaz FA 2000, Biotic invasions: causes, epidemiology, global consequences, and control. *Ecol. Appl.* **10**: 689-710.

[4] Boudouresque CF, and Verlaque M 2002, Assessing scale and impact of ship-transported alien macrophytes in the Mediterranean Sea. In Alien marine organisms introduced by ships in the Mediterranean and Black seas. F. Briand ed. CIESM Workshop Monographs 20, Monaco: 53-61.

[5] Baffreau A, Pezy JP, Rusig AM, Mussio I, and Dauvin JC 2018, Les espèces marines animales et végétales introduite en Normandie. Imprimerie Moderne de Bayeux, 350 pp.

[6] Seebens H, Blackburn T, Dyer E et al. 2017, No saturation in the accumulation of alien species worldwide. *Nat. Commun.* **8**: 14435.

[7] Afonso I, Berecibar E, Castro N, Costa JL, Frias P, Henriques F, Moreira P, Oliveira PM, Silva G, and Chainho P 2020, Assessment of the colonization and dispersal success of non-indigenous species introduced in recreational marinas along the estuarine gradient. *Ecol. Ind.* **113**: 106-147.

[8] Ros M, Vázquez-Luis M, and Guerra-García JM 2013, The role of marinas and recreational boating in the occurrence and distribution of exotic caprellids (Crustacea: Amphipoda) in the Western Mediterranean: Mallorca Island as a case study. *J. Sea Res.* **83**: 94-103.

[9] Chainho P, Fernandes A, Amorim A, Ávila SP, Canning-Clode J, Castro JJ, Costa AC, Costa JL, Cruz T, Gollasch S, Grazziotin-Soares C, Melo R, Micael J, Parente MI, Semedo J, Silva T, Sobral D, Sousa M, Torres P, Veloso V, and Costa MJ 2015, Non-indigenous species in Portuguese coastal areas, coastal lagoons, estuaries and islands. *Estuar. Coast. Shelf Sci.* **167**: 199-201.

[10] Martínez-Laiz G, Ulman A, Ros M, and Marchini A 2019, Is recreational boating a potential vector for non-indigenous peracarid crustaceans in the Mediterranean Sea? A combined biological and social approach. *Mar. Poll. Bull.* **140**: 403-414.

From concretes to bioreceptive concretes, influence of concrete properties on the biological colonization of marine artificial structures

M Hayek[1], M Salgues[1], J C Souche[1], K D Weerdt[2] and S Pioch[3]

[1] LMGC, IMT Mines Alès, Univ Montpellier, CNRS, 6, avenue de Clavières, 30319 Alès Cedex, France
[2] Norwegian University of Science and Technology NTNU, Department of Structural Engineering, Trondheim, Norway
[3] Université Paul Valéry Montpellier 3, Site Saint Charles, Laboratoire de Géographie et d'Aménagement de Montpellier (LAGAM), Rue du Professeur Henri Serre, 34080 – Montpellier

Abstract. A main cause of biodiversity loss is artificialization of the marine environment (IPBES 2019). With 39,400 km² of coastal and marine areas already encroached upon by human infrastructure and an increasing demand on space due to the growing global population – projected to reach 9 billion by 2050 – it is clear that humanity needs to find ways to prevent its activities from endangering biodiversity. To this end, since the 1990s ecologists have been trying to develop a win-win approach that unites ecological engineering with civil engineering. Today, civil engineers have a responsibility to incorporate eco-design processes in all construction projects underway to ensure that the latter benefit both humans and nature. Then, the new challenge of the 21st century is to develop eco-designed concretes that, in addition to their usual properties, provide improved bioreceptivity in order to enhance marine biodiversity without affecting the structure durability. The aim of this study is to master, clarify and classify the intrinsic parameters that influence the bioreceptivity (biocolonization) of cementitious materials in the marine environment. By using biofilm-culture-method (biofilm quantification), this study shows that the use of rough surface or slag cement CEM III and the surface treatment with green formwork oil enhance the biocolonization of cementitious materials in the marine environment whereas the application of curing agent (hydrophobic surface coating) has the opposite effect. Among the influent parameters, surface roughness proved to be the factor that promotes biocolonization most effectively.

Key words: Eco-design; cementitious materials; marine environment; Intrinsic parameters; Biocolonization, Bioreceptivity

1. Introduction

In the marine environment, microbial cells are often found in complex, surface-attached communities, known as biofilms. By definition, biofilms are assemblages of single or multiple microbial populations irreversibly attached to a substrate or an interface and included in a coating of extracellular polymeric substances (EPS) [1]. Biofilms arbor considerable microbial diversity. Bacteria, archaea, algae, fungi, protozoa and viruses all form important components of the biofilm matrix and dominate biogeochemical and bioremediation processes in seawater [2,3]. Then, biofilms contribute to the biodiversity and ecosystem processes of aquatic ecosystems.

Moreover, any surface in the marine environment will be colonized with micro- and macro-organisms leading to biofilm formation [4]. This biocolonization process develops via four distinct phases; adsorption of dissolved organic molecules, attachment of bacterial cells, attachment of unicellular eukaryotes and attachment of larvae and spores [5]. Bacterial attachment is a highly controlled and regulated process whereby attached cells produce EPS to form structured and complex matrixes [6]. Then, these pioneer microorganisms facilitate the establishment of the next arrivals, including heterotrophic and photosynthetic microorganisms such as cyanobacteria, fungi, diatoms, barnacles, algae and protozoa [7].
However, in alignment with the global attention to environmental protection and biodiversity restoration in the marine environment [8], the number of studies about the biocolonization of submerged materials and their effects on marine ecology and biodiversity has increased in the last decade. Indeed, in the marine environment, urban, coastal and offshore structures provide an important protective function, but can also have unintended ecological consequences, such as the loss or modification of habitat and the alteration of hydrological and ecological flows [9,10]. The new challenge of the 21st century for managers is to eco-design marine infrastructure in a way that minimizes its ecological impact and increases its bioreceptivity (the ability to be colonized by living organisms) in order to enhance marine biodiversity without losing the structure durability [11]. Knowing that concrete is one of the materials most widely used for the construction of marine infrastructure such as ports and coastal defences [12], master and clarify the intrinsic parameters that influence the bioreceptivity of cementitious materials in the marine environment are crucial and constitute a fundamental step towards more green and eco-designed marine artificial structures.
However, the durability of a concrete structure in the marine environment is influenced by the chemical, physical and biological actions [13]. Influences of a physical and chemical nature are generally well studied, and are subject to standards and recommendations [14]. Several scientific publications state that actions of a chemical nature such as chloride and magnesium sulfate attack are the main cause of concrete structure deterioration in the marine environment [13,15]. However, the actions of a biological nature (e.g., colonization of concrete by marine organisms) have been less studied, and less is known about them [14]. The effect of these actions on the durability of concrete in the marine environment remains unclear, but most scientists agree that marine organisms adhered to the concrete surface have a protective effect (bioprotection) against chemical attack in seawater [16–18]; they form a physical barrier that reduces surface permeability. The decrease in surface permeability leads to less-efficient diffusion of aggressive ions (Cl^-, Mg^{2+} and OH^-), which can increase the durability of a concrete structure in the marine environment [19,20].
Generally speaking, the intrinsic parameters that influence the biocolonization (biofilm formation) of cementitious materials are the nature and the physicochemical properties of the surface: chemical composition, roughness, porosity, hydrophobicity and pH [21–27]. However, these factors are less well known in the marine environment. We showed in a previous study that the bacterial colonization of cementitious materials in seawater is influenced by the pH and the type of cement [28,29].
In order to help engineers in their new challenge of designing marine infrastructures using eco-designed concrete, the main objective of this present study is to determine the effect of some intrinsic parameters on the biocolonization of cementitious materials in the marine environment. These parameters are the the use of curing compound or green formwork oil, the surface roughness and the type of cement.

2. Materials and Methods

2.1. Preparation of Mortar Specimens

Ordinary Portland Cement (OPC) CEM I cement 52.5 N CE CP2 NF “SB” (provided by Ciments Calcia) and blastfurnace slag cement CEM III (composed of 60% ground granulated blast-furnace slag NF EN 15167-1, provided by ECOCEM (CAS no.: 65996-69-2) were used in this study to produce six types of mortar specimens (Table 1). The mortar had a water/cement ratio (w/c) of 0.5 and was composed of 450 g cement and 1350 g sand (sand 0/4 EN 196-1).

Table 1. Types and compositions of mortar specimens investigated in this study.

Mortar Specimen		Surface Type	Cement	CEM I (g)	CEM III (g)	Water (g)	Sand (g)	Curing Agent	Formwork Oil
Control mortar	Ref	Smooth (SS)	CEM I	450	0	225	1350	-	-
Cured mortar	CM		CEM I	450	0	225	1350	+	-
Oiled mortar	OM1		CEM I	450	0	225	1350	-	+
	OM3		CEM III	180	270	225	1350	-	+
Biomimetic mortar	BM1	Rough (RS)	CEM I	450	0	225	1350	-	-
	BM3		CEM III	180	270	225	1350	-	-

After mixing, the mortars were cast in cylindrical molds measuring 2.7 cm in diameter and 2.9 cm high. After 24 h of hardening, the mortar specimens were removed from the molds and placed in a laboratory room at 20 °C for 30 days. Formwork oil (vegetable oil, BIODEM SI 3) was spread on the molds, using absorbent paper to avoid any surplus, before the mortar was poured. Curing compound (SikaCem®Cure) was added to the top of the specimens in accordance with the supplier's instructions 1 h after stripping. The rough mortars were prepared with a rough silicone skin; no release agent was needed. The cylindrical molds filled with mortar were poured on the silicone skin and held in place with a weight.

2.2. *Immersion Conditions*

An immersion test in seawater was carried out using flat-bottomed basins (polyester, 6 m long, 0.6 m high and 2 m wide) located at the IFREMER station (biology of exploited marine organism's research unit in Palavas, France). The basins featured a seawater inlet and outlet allowing for an open seawater circuit. To ensure that the experiment would be completed smoothly and avoid contamination of any type, the basins were cleaned before the cementitious material specimens were placed in them.

2.3. *Quantification of bacterial colonization*

Quantification of the bacterial biofilm adhered to the surface of the mortar specimens was performed using the protocol described by Hayek *et al.*, (culture-based methods) [28]. This quantification was carried out after 0, 1, 2, 6, 8, 15, 24, 26 and 28 days of immersion in seawater. After each incubation period, three specimens of each mortar type were placed in three sterile tubes containing 10 mL of sterile seawater. Bacteria adhering to the mortar surface were detached by immersing the tubes in an ultrasonic bath (Bandelin SONOREX™) for 10 min at 20 °C. The solution obtained was diluted using sterile seawater and 100 µl of diluted solution were spread on plates containing marine agar (MA, Dutscher, 490614). The plates were then incubated at 20 °C, and a colony count was performed at least 72 h after incubation. The results are expressed as colony-forming units per cm^3 of mortar (CFU/cm3).

2.4. *Hydrophobicity Evaluation*

The hydrophobicity of the mortar surface during bacterial colonization was evaluated using a drop-shape scanner (KRÜSS, DSA 30), which measures the angles and diameters of contact between a drop of water and a surface [30]. After each period of incubation in seawater, 3 mortar specimens were drained for 2 h. Then, a drop (3 µL) of water was placed on mortar surface. Absorption of the drop was monitored using a camera (8 images per second). The contact angle between the drop and the material was measured using the Advance machine software, and a minimum of three tests were carried out at different points.

2.5. Statistical Analyses
In order to evaluate the significance of the various biocolonization obtained, statistical analysis was performed using GraphPad Prism 5 (GraphPad Software, San Diego, CA, USA) and one-way ANOVA tests. Statistical significance was accepted by p-value <0.05 obtained using Bonferroni or Tukey's multiple comparison post hoc tests.

3. Results and discussion
The biological colonization of cementitious materials is affected according to the literature by many factors, including the environmental conditions, the bacterial properties and the physical/chemical characteristics of the material surface [24,31]. To avoid the effect of the environmental conditions and bacterial parameters on the biocolonization results, the mortar specimens were incubated at the same time under the same environmental conditions in the presence of the same marine bacteria. Therefore, only the physicochemical properties of the mortar surfaces could generate a different rate of bacterial colonization.

3.1. Effect of the curing compound and the green formwork oil on the biocolonization of mortar specimens
Generally, the curing compound and the green formwork oil are used during manufacturing to improve the durability of concrete; the curing agent is a liquid applied to the concrete or mortar surface to protect the material from water evaporation and give it greater aesthetic and mechanical durability, preventing early-age surface cracking [32]. Formwork oil is a mold-release agent that is applied to a wall of a mold to ensure easy separation of the hardened concrete from the mold by reducing the adhesion forces at the concrete/mold interface [33]. This oil forms a well-adsorbed and stable "lubricant monolayer" on the surface of cementitious materials, leading to improved release performance [34].
The effect of this tow products (curing compound and green formwork oil) is poorly understood to date. Figure 1 shows that the curing compound inhibits the bacterial colonization of mortar submerged in seawater whereas green formwork oil has the opposite effect.

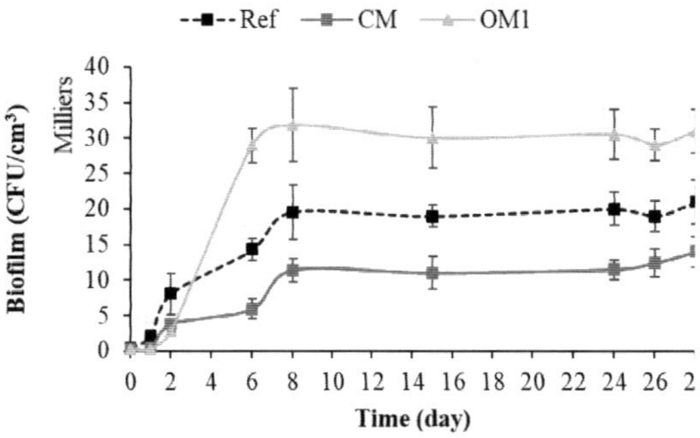

Figure 1. Quantification of bacterial colonization of mortar specimens. Each quantification was performed in triplicate using the culture-based method, and the error bars present the standard deviation from the values obtained. Ref = control mortar; CM = cured mortar; OM1 = oiled mortar prepared with CEM I.

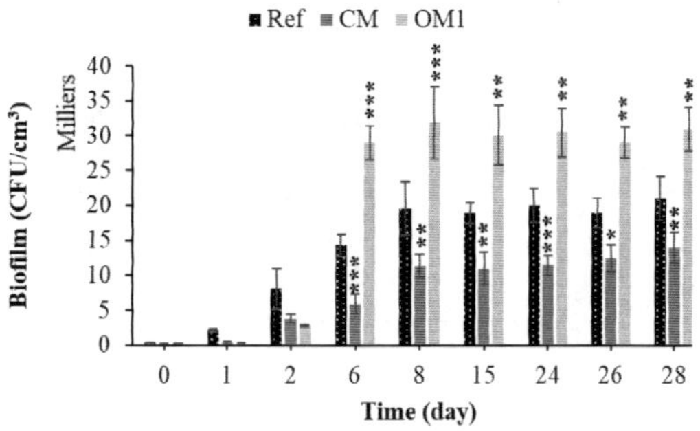

Figure 2. Statistical comparison (GraphPad Prism 5) of the biocolonization of mortar surfaces treated and untreated with curing compound or green formwork oil. Each experiment was performed in triplicate, and the error bars present the standard deviation of the values obtained. The experiments highlighted with asterisks differed significantly from the control (Bonferroni; **: p < 0.01, ***: p < 0.001) at the indicated time.

Figure 1 shows that the bacterial colonization of mortar surfaces started with a latency phase, followed by a phase of cell growth and accumulation on the surfaces and ending with a plateau phase. These

colonization kinetics were also observed in most of the studies quantifying bacterial-biofilm formation on cementitious-material surfaces or on other surface types [27,28].
Figure 2 shows that the curing compound significantly inhibits the bacterial colonization of mortar submerged in seawater. Cells accumulated and grew faster and more extensively on an untreated surface. The compositions of curing compound used in this study indicate the presence of alkyl (C14-C18) bis (2-hydroxyethyl) amine, 5-chloro-2-methyl-2H-isothiazol-3-one and 2-methyl-2H-isothiazol-3-one (FDS Sikacem cure). These elements or their derivatives have been cited as an anti-biofilm molecules that inhibited the formation and the accumulation of bacterial biofilm [35,36]. Moreover, the treatment of the material by an hydrophobic surface coating is cited as one of the anti-biofilm strategies used to inhibit bacterial and other organisms adhesion on the surface [37–39]. To verify the effect of the curing film on the hydrophobicity of the mortar surface, the contact angle with a drop of water was measured for the treated and untreated surface using a drop-shape scanner. From their preparation until the start of the immersion test, the mortars treated with the curing compound presented a contact angle greater than 110° (40° in the case of untreated mortars), indicating a hydrophobic surface [40]. After immersion and under the action of seawater, the contact angle gradually decreased with time and became equal to that of the control mortar at 28 days of immersion (data not shown). Therefore, we propose that the curing compound inhibited the biological colonization of mortar surfaces in seawater because of its chemical composition (anti-biofilm) and its effect on surface hydrophobicity.
However, Figure 2 shows that the formwork oil significantly promotes the bacterial colonization of mortar submerged in seawater. The composition of the green formwork oil used in this study is not given. Detailed information is given neither in the technical product information nor in the bibliography. Therefore, knowing that green formwork oil is biodegradable, we propose that this oil applied on the surface of the mortar specimens was used as a carbon source by marine bacteria, according to the study and the results of [41,42]. Dusane *et al.* showed by a laboratory test that the biofilm formation is affected by the carbon sources. Lactic acid, erythritol, glycerol, glucose and edible oils increase this process [43]. Therefore, we propose that the surface treatment with this type of formwork oil increased the biological colonization of cementitious materials in the marine environment because it is used as a nutrients source by marine microorganisms.

3.2. Effect of the type of cement and the surface roughness on the biocolonization of mortar specimens.

In order to compare and classify the intrinsic parameters that influence the biocolonization of cementitious materials in the marine environment, we also tested in this study under the same conditions the effect of the type of cement (the use of CEM III) and the surface roughness on the biofilm formation at the surface of mortar specimens.

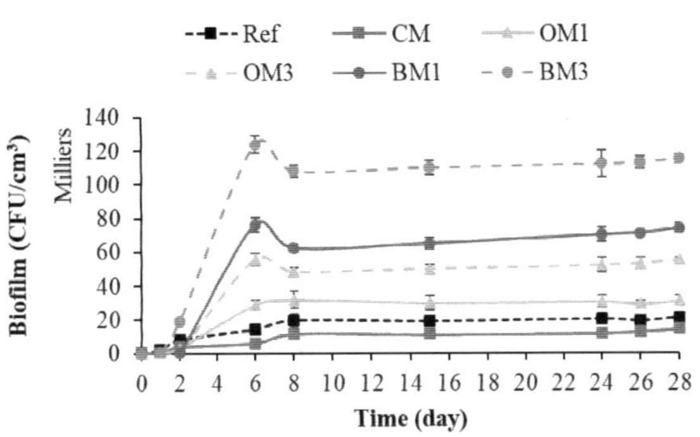

Figure 3. Quantification of bacterial-biofilm formation on mortar specimens. Each quantification was performed in triplicate using the culture-based method, and the error bars present the standard deviation from the values obtained. Ref = control mortar; CM = cured mortar; OM1 = oiled mortar prepared with CEM I; OM3 = oiled mortar prepared with CEM III; BM1 = biomimetic mortar prepared with CEM I; BM3 = biomimetic mortar prepared with CEM III.

Figure 3 shows that the bacterial colonization of mortar surfaces in the case of OM3, BM1 and BM3 has the same kinetic observed in the case of Ref, CM and OM1 (cf. Figure 1). However, the use of CEM III (BM1 vs. BM3 or OM1 vs. OM3) significantly increases the biofilm gormation. These results

were in keeping with the literature, in which a similar effect of cement type on the biological colonization of cementitious materials has been reported [28,44,45]. In addition, a rough mortar surface significantly increased bacterial colonization in comparison with a smooth surface (Ref). This influence of surface roughness was identified in several studies concerning the biological colonization of cementitious materials [27,44].
Therefore, based on the rate of biological colonization (Figure 3), the mortar types can be classified from less to more bioreceptive in the following order: CM< Ref < OM1 < OM3 < BM1 < BM3. Then, the intrinsic parameters that promotes the biocolonization of cementitious materials in the marine environment can be classified from more to less effectively in the following order: surface roughness > the use of CEM III (type of cement) > surface treatment with green formwork oil.
However, the main cause of concrete structure deterioration in the marine environment is the chemical attack caused by seawater ions such as chloride and magnesium sulfate attack [13]. According to the literature, the biological colonization of immerged concrete structure can have a protective effect (bioprotection) against chemical attack in seawater, leading to improved structure durability [19,20]; the marine organisms adhered to the concrete surface form a physical barrier that reduces surface permeability. The decrease in surface permeability leads to less-efficient diffusion of aggressive ions (Cl^-, Mg2+ and OH^-), which can increase the durability of a concrete structure in the marine environment. Furthmore, Lv *et al.* demonstrate that the presence of the *Crassostrea gigas* coating (biological coating) on the surface of marine concrete can reduce the water absorption of concrete, and enhance the resistance of concrete to chloride penetration and carbonation [46]. Then, improving the biological colonization of concrete structure in the marine environment can have a positive effect both for the environment (biodiversity restoration) and for the concrete structure (bioprotection).

4. Conclusions

Eco-design of marine structures is a major focus for many researchers and construction companies working in marine environment, to enhance durability and also since few years to minimize and mitigate human impacts toward "no net loss" on biodiversity policies **[26]**. This study tests, compares and classify the intrinsic parameters that influence the biocolonization of cementitious materials in the marine environment. Regarding the parameters tested in the present work, we can summarize that: (i) work practices such as the use of a curing agent and/or formwork oil have an impact on biocolonization; the surface treatment with green formwork oil enhance the biocolonization whereas the application of curing agent has the opposite effect. (ii) The use of rough surface or slag cement CEM III increases the bioreceptivity of cementitious materials. (iii) Among the parameters examined, surface roughness proved to be the factor that promotes biocolonization most effectively.
These results could be taken up in future recommendations to enable engineers to eco-design more eco-friendly marine infrastructure and develop green-engineering projects.

Acknowledgments

Authors wishing to acknowledge the financial support from ARTELIA. The authors thank Emmanuel Rezzouk (IFREMER, Biology Research Unit for exploited marine organisms) for receiving them at the Palavas-les-Flots oceanographic station and thank LIB industry for the supply of the silicone matrix and their assistance.

References

[1] Dobretsov S, Abed R M M and Teplitski M 2013 Mini-review: Inhibition of biofouling by marine microorganisms *Biofouling* **29** 423–41
[2] Besemer K 2015 Biodiversity, community structure and function of biofilms in stream ecosystems *Research in Microbiology* **166** 774–81
[3] Singh R, Paul D and Jain R K 2006 Biofilms: implications in bioremediation *Trends in Microbiology* **14** 389–97

[4] Flemming H-C 2002 Biofouling in water systems – cases, causes and countermeasures *Appl Microbiol Biotechnol* **59** 629–40
[5] Lobelle D and Cunliffe M 2011 Early microbial biofilm formation on marine plastic debris *Marine Pollution Bulletin* **62** 197–200
[6] Donlan R M 2002 Biofilms: Microbial Life on Surfaces *Emerg Infect Dis* **8** 881–90
[7] Grzegorczyk M, Pogorzelski S J, Pospiech A and Boniewicz-Szmyt K 2018 Monitoring of Marine Biofilm Formation Dynamics at Submerged Solid Surfaces With Multitechnique Sensors *Front. Mar. Sci.* **5**
[8] Bai X, Geschke A, Molnár Z, Jaureguiberry P, Hendry A, Liu J, Martin A, Midgley G F, Scheffran J, Seppelt R, Strassburg B, Spangenberg J, Stenseke M, Turnhout E, Williams M J and Zayas C 2019 IPBES Global Assessment on Biodiversity and Ecosystem Services Chapter 1. Assessing a planet in transformation: Rationale and approach of the IPBES Global Assessment on Biodiversity and Ecosystem Services 69
[9] Heery E C, Bishop M J, Critchley L P, Bugnot A B, Airoldi L, Mayer-Pinto M, Sheehan E V, Coleman R A, Loke L H L, Johnston E L, Komyakova V, Morris R L, Strain E M A, Naylor L A and Dafforn K A 2017 Identifying the consequences of ocean sprawl for sedimentary habitats *Journal of Experimental Marine Biology and Ecology* **492** 31–48
[10] Waltham N J and Dafforn K A 2018 Ecological engineering in the coastal seascape *Ecological Engineering* **120** 554–9
[11] Browne M A and Chapman M G 2011 Ecologically Informed Engineering Reduces Loss of Intertidal Biodiversity on Artificial Shorelines *Environ. Sci. Technol.* **45** 8204–7
[12] Khanzadeh Moradllo M, Shekarchi M and Hoseini M 2012 Time-dependent performance of concrete surface coatings in tidal zone of marine environment *Construction and Building Materials* **30** 198–205
[13] Yi Y, Zhu D, Guo S, Zhang Z and Shi C 2020 A review on the deterioration and approaches to enhance the durability of concrete in the marine environment *Cement and Concrete Composites* 103695
[14] Jacob C, Buffard A, Pioch S and Thorin S 2018 Marine ecosystem restoration and biodiversity offset *Ecological Engineering* **120** 585–594
[15] Costa A and Appleton J 2002 Case studies of concrete deterioration in a marine environment in Portugal *Cement and Concrete Composites* **24** 169–79
[16] Pioch S, Relini G, Souche J C, Stive M J F, De Monbrison D, Nassif S, Simard F, Allemand D, Saussol P, Spieler R and Kilfoyle K 2018 Enhancing eco-engineering of coastal infrastructure with eco-design: Moving from mitigation to integration *Ecological Engineering* **120** 574–84
[17] Chlayon T, Iwanami M and Chijiwa N 2018 Combined protective action of barnacles and biofilm on concrete surface in intertidal areas *Construction and Building Materials* **179** 477–87
[18] Gowell M R, Coombes M A and Viles H A 2015 Rock-protecting seaweed? Experimental evidence of bioprotection in the intertidal zone *Earth Surface Processes and Landforms* **40** 1364–70
[19] Soleimani S, Ormeci B and Isgor O B 2013 Growth and characterization of Escherichia coli DH5α biofilm on concrete surfaces as a protective layer against microbiologically influenced concrete deterioration (MICD) *Applied microbiology and biotechnology* **97** 1093–1102
[20] Perkol-Finkel S and Sella I 2014 Ecologically Active Concrete for Coastal and Marine Infrastructure: Innovative Matrices and Designs *From Sea to Shore? Meeting the Challenges of the Sea* vol 1 and 2 (ICE Publishing) pp 1139–49
[21] Lors C, Feugeas F and Tribollet B 2019 *Interactions Materials - Microorganisms: Concrete and Metals more Resistant to Biodeterioration* (EDP sciences)

[22] Dubosc A, Escadeillas G and Blanc P J 2001 Characterization of biological stains on external concrete walls and influence of concrete as underlying material *Cement and Concrete Research* **31** 1613–7
[23] Manso S, De Muynck W, Segura I, Aguado A, Steppe K, Boon N and De Belie N 2014 Bioreceptivity evaluation of cementitious materials designed to stimulate biological growth *Science of The Total Environment* **481** 232–41
[24] Hayek M, Salgues M, Habouzit F, Bayle S, Souche J-C, Weerdt K D and Pioch S 2020 La bioréceptivité de matériaux cimentaires dans l'eau de mer : mécanismes, facteurs agissants et conséquences *Revue Paralia* **13**
[25] Souche J-C, Pioch S, Salgues M, Weerdt K D, Agostini A and Hayek M 2019 De la conception à l'éco-conception des ouvrages maritimes : intégrer la nature au projet d'aménagement maritime *Revue Paralia* **12**
[26] Salgues M, Pioch S, Souche J-C and Weerdt K D 2020 L'écoconception maritime : une révolution pour les maitres d'ouvrages *Revue Paralia* **13**
[27] Tran T H, Govin A, Guyonnet R, Grosseau P, Lors C, Damidot D, Deves O and Ruot B 2014 Influence of the intrinsic characteristics of mortars on their biofouling by pigmented organisms: Comparison between laboratory and field-scale experiments *International Biodeterioration & Biodegradation* **86** 334–42
[28] Hayek M, Salgues M, Habouzit F, Bayle S, Souche J-C, De Weerdt K and Pioch S 2020 In vitro and in situ tests to evaluate the bacterial colonization of cementitious materials in the marine environment *Cement and Concrete Composites* **113** 103748
[29] Hayek M, Salgues M, Habouzit F, Bayle S, Souche J-C, Weerdt K D and Pioch S 2020 L'influence de la carbonatation sur la biocolonisation de matériaux cimentaires dans le milieu marin *Matériaux & Techniques* **108** 202
[30] Munzer C, Belhaj E, Meylheuc T, Lecomte A and Feugeas F 2015 Effets d'un bioadjuvant sur les caractéristiques de surface de pâtes cimentaires *Matériaux & Techniques* **103** 208
[31] S. Karthick Raja Namasivayam, Francis A L, Arvind Bharani R S and Nachiyar C V 2019 Bacterial biofilm or biofouling networks with numerous resilience factors from real water supplies of Chennai and their enhanced susceptibility to biocompatible nanoparticles *Journal of Cleaner Production* **231** 872–98
[32] Dang Y, Qian J, Qu Y, Zhang L, Wang Z, Qiao D and Jia X 2013 Curing cement concrete by using shrinkage reducing admixture and curing compound *Construction and Building Materials* **48** 992–7
[33] Djelal C, Vanhove Y, De Caro P and Magnin A 2002 Role of demoulding agents during self-compacting concrete casting in formwork *Mat. Struct.* **35** 470–6
[34] León-Martínez F M, Abad-Zarate E F, Lagunez-Rivera L and Cano-Barrita P F de J 2016 Laboratory and field performance of biodegradable release agents for hydraulic concrete *Mater Struct* **49** 2731–48
[35] Andjouh S and Blache Y 2016 Screening of bromotyramine analogues as antifouling compounds against marine bacteria *Biofouling* **32** 871–81
[36] Andjouh S and Blache Y 2015 Click-based synthesis of bromotyrosine alkaloid analogs as potential anti-biofilm leads for SAR studies *Bioorganic & Medicinal Chemistry Letters* **25** 5762–6
[37] Huggett M J, Nedved B T and Hadfield M G 2009 Effects of initial surface wettability on biofilm formation and subsequent settlement of Hydroides elegans *Biofouling* **25** 387–99
[38] Blainey B L and Marshall K C 1991 The use of block copolymers to inhibit bacterial adhesion and biofilm formation on hydrophobic surfaces in marine habitats *Biofouling* **4** 309–18
[39] Razavi S M R, Oh J, Haasch R T, Kim K, Masoomi M, Bagheri R, Slauch J M and Miljkovic N 2019 Environment-Friendly Antibiofouling Superhydrophobic Coatings *ACS Sustainable Chem. Eng.* **7** 14509–20

[40] Bixler G D and Bhushan B 2012 Biofouling: lessons from nature *Philosophical Transactions of the Royal Society A: Mathematical, Physical and Engineering Sciences* **370** 2381–417

[41] Li Z, Wrenn B A and Venosa A D 2005 Anaerobic biodegradation of vegetable oil and its metabolic intermediates in oil-enriched freshwater sediments *Biodegradation* **16** 341–52

[42] Mehdi H and Giti E 2008 Investigation of alkane biodegradation using the microtiter plate method and correlation between biofilm formation, biosurfactant production and crude oil biodegradation *International Biodeterioration & Biodegradation* **62** 170–8

[43] Dusane D H, Nancharaiah Y V, Venugopalan V P, Kumar A R and Zinjarde S S 2008 Biofilm formation by a biotechnologically important tropical marine yeast isolate, Yarrowia lipolytica NCIM 3589 *Water Science and Technology* **58** 1221–9

[44] Voegel C, Durban N, Bertron A, Landon Y and Erable B 2020 Evaluation of microbial proliferation on cementitious materials exposed to biogas systems *null* **41** 2439–49

[45] Ly O, Yoris-Nobile A I, Sebaibi N, Blanco-Fernandez E, Boutouil M, Castro-Fresno D, Hall A E, Herbert R J H, Deboucha W, Reis B, Franco J N, Teresa Borges M, Sousa-Pinto I, van der Linden P and Stafford R 2021 Optimisation of 3D printed concrete for artificial reefs: Biofouling and mechanical analysis *Construction and Building Materials* **272** 121649

[46] Lv J, Cao Z and Hu X 2021 Effect of biological coating (Crassostrea gigas) on marine concrete: Enhanced durability and mechanisms *Construction and Building Materials* **285** 122914

IOP Conf. Series: Materials Science and Engineering 1245 (2022) 012009 doi:10.1088/1757-899X/1245/1/012009

The involvement of actors in the English Channel through a socio-ecological approach applied to artificial reefs

J Salaün[1,2,3], S Pioch[2], J C Dauvin[3]
[1] *Atlantique Landes Récifs, 1 rue du Pitey, ZA des deux pins, 40130 Capbreton*
[2] *Université Paul Valéry, Montpellier 3, Laboratoire d'Aménagement et de Géographie de Montpellier EA LAGAM, F34000, Montpellier, France.*
[3] *Normandie Univ., UNICAEN, CNRS, UMR 6143 M2C, Laboratoire Morphodynamique Continentale et Côtière, 24 rue des Tilleuls, 14000 Caen, France*

Abstract. For more than fifty years, Artificial Reefs (AR) have been deployed in France to respond to the decline in fish stocks with a high deployment on the Mediterranean coast. However, from 40 sites listed, less than a quarter have not been the subject of any known published ecological or socio-economic monitoring reports, over the past five years. The lack of data on the evaluation of these structures therefore raises questions about their interest, both from an ecological and social point of view. By relying on the theory of organizations and more specifically on the translation resulting from the sociology of innovation applied to the environment, we will develop an original socio-ecological approach intended to meet this need for evaluation, in particular for managers. Based on two case studies on the coasts of the French Channel (Etretat and Cherbourg), we will highlight the interests and roles of human and non-human actors in an AR immersion process. The analysis will highlight the social and ecological objectives of these two projects from an analytical framework designed to verify the adequacy between the socio-ecological network "artificial reefs" formed and the objectives initially expected by the actors, especially decision-makers and funders. The limits posed by these approaches to improve the evaluation of artificial reefs will finally be discussed.

1. Introduction

Due to overexploitation and decrease in fish stocks [1], new tools for the management of the coastal zone have emerged around the worldwide ocean. Artificial reefs (ARs) are human structures voluntarily deployed for the objectives of fixing, concentrating, increasing or regenerating marine fauna [2]. They are used in many countries around the world for the purposes of fish production (such as in Japan, China or Portugal), development of recreational activities (such as in Australia, United States or Israel) or protection against illegal trawling actions (such as in France, Spain or Italy). Scientific research on ARs applies to these three main objectives.

In France, since the 1960s, ARs deployment have increased considerably. On all French metropolitan coastlines, the number of ARs sites reaches 44 in 2020. Historically, the main objective of French ARs is to support the production of artisanal fisheries. Indeed, in the Mediterranean Sea, over 2/3 of the ARs sites have a fisheries production objective [3]. However, the distribution of the number of ARs sites between French maritime coast is very heterogeneous. When the Mediterranean Sea has about 30 ARs sites in 2020, North Brittany and the English Channel have only four ARs sites and the Bay of Biscay nine.

For more than 50 years that ARs have been used in France, their assessments remain unusual or incomplete [2 -7]. The studies carried out often target indicators such as the abundance and the species richness of or focus only on a specific fish community or on some species [8]. These studies, although they have shown the direct effect of ARs on the increase of local primary production [9 -10], provide only a partial answer to the problem of assessing the contribution of ARs within a socio-ecosystem point of view. Indeed, the socio-economic dimension is often forgotten [11 -12]. To compensate this lack, some studies have focused on landing data, experimental fisheries, calculations of biological production of ARs from the selling prices of species at the harbour markets. Others have integrated the social dimension by studying users' perception of the usefulness of ARs [6;13]. Nevertheless, no study considered the entire socio-ecological network to assess the overall contributions of ARs.

A socio-ecological approach applied to ARs sites seems relevant to understand the overall functioning of the ARs system but also to compare the social-ecological contributions of ARs with the initial expectations of the actors of the territory. The use of this innovative approach is part of a multidisciplinary approach linking nature and culture within a territory.

Based on two case studies on the French English Channel (Etretat and Cherbourg), we highlight in this paper the interests and roles of human and non-human actors in an ARs project.

2. Materials and Methods

2.1. Study sites

The study sites chosen are located in the English Channel, off Etretat and in the harbour of Cherbourg (Figure 1).

In 2008, 450 m^3 of the ARs were deployed off Etretat by the Chamber of Commerce and Industry (CCI) of Fécamp/Bolbec. The 25-ha site is located 2.4 km from the coast and on depths of -17m. It is composed of a central module of 54 m^3, a first belt of heap of cubic reefs and a second belt of protective reefs. The concrete used is composed of synthetic fibber that favours fauna and flora colonization. The project was financed by European funds dedicated to fisheries, the Haute-Normandie Region and the Seine-Maritime Department. The objective of the ARs site is to support the fishing industry by targeting species with fisheries interest such as sea bass [14].

The second study site is located in the Cherbourg harbour on shallow depths up to 5m deep. The ARs were deployed in 2015 and was design during the European RECIF project. The modules are composed of 72 experimental shell concrete blocks, spread over three levels of 24 units for a total volume of approximatively 100 m^3. The objective is to improve the biodiversity and production of the English Channel marine ecosystems. The RECIF project was selected within the framework of the European Cross-border Cooperation Programme INTERREG IV A France (English Channel) / England, co-financed by the European Regional Development Fund (ERDF). It brings together eight French and British partners and was supported by the 'Conseil Régional de Basse-Normandie'. The RECIF project is part of the opportunities and actions carried out for the improvement of the ecosystem of the English Channel and for a better management of marine resources [15].

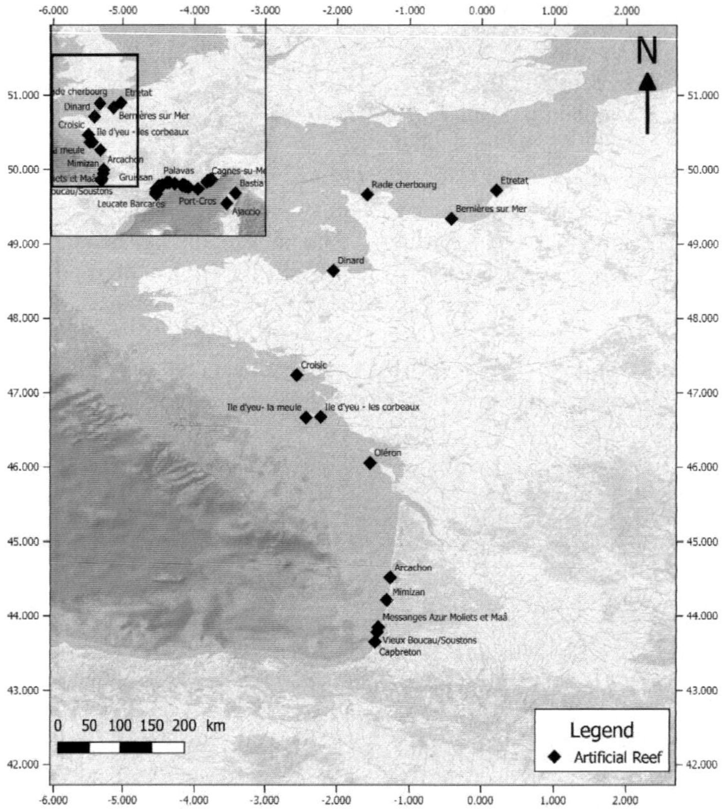

Figure 1. Location map of the artificial reef sites in the English Channel

2.2. The Translation concept

The concept of translation had been developed by Callon [16] and Latour [17] and based on the Actor Network Theory [18]. This concept is used to describe each step of the actor network construction to understand the process of collective action [17]. The particularity of this concept is to consider the non-human actors who can be both technical objects and living beings as part of the definition of actors. This particularity makes it possible to consider marine fauna and flora as an actor in the ARs socio-ecosystem analysis.

The translation concept is defined by four steps [16;19;20]:

1. Problematization, which defines the issues and the obligatory passage point for collective action (the common solution responding to the challenges identified);
2. Interessement, which expresses the interests of each actor;
3. Enrolment, which assigns roles to each actor;
4. Mobilization of allies, which makes it possible to extend the network and gather other actors to collective action.

2.3. Data acquisition

In order to understand how the ARs network was created, qualitative open-ended interviews were conducted. This methodology does not seek to cover representativeness but the diversity and complexity of social reality [21]. During the first interview campaigns, a model is built and is constantly modified by the contribution of the following interviews. Then the model is consolidated until the new interviews no longer provide new information. The model then called "saturated", validates the results obtained [22].

The choice of the actors interviewed follows the "snowball effect" method which consists of starting from a central actor of the study case, in our case the ARs site manager, then all the actors mentioned

during the interview of the central actor. These actors identified (level 1) directly by the central actor are also interviewed and the new actors mentioned are integrated in the network (level 2).
Human actors have been grouped defining three categories of actors according to their social function [23]:
-The State;
-Local government;
-Civil society.
The non-human actors, who could not be questioned directly, were represented by scientists as experts of the marine fauna.

Table 1. Number and type of actors interviewed

Type of actors interviewed	Number
State	2
Local government	1
Civil society	12
Wildlife Representative	6

3. Results

3.1. A favourable socio economic context to deploy artificial reefs

During the 1990-2000, ARs is democratized in the Mediterranean Sea and the Atlantic coast. Many projects were born during this period.
The materials used are mainly recycled elements such as electric poles, nozzles, blocks... Since 1999, ARs projects are eligible to receive European funds in France. This financial assistance, along with the reduction in costs through the use of recycled materials, makes ARs projects more affordable for local government such as townships. Acceptance of ARs is also social. At this time, the fisheries sector is facing a decrease of fish catches [24]. In consequence, the professional fishermen, seeing in this tool a manner to fix the resource and manage conflicts between inshore fishermen and trawlers, requested ARs deployment. The media also take up the subject and relay the deployment through reports on national TV channels such as Thalassa. From a scientific point of view, after an experiment carried out in 1990s, the reports conclude that it is difficult to assess the real benefits for the fisheries sector [5;25].
This socio-economic context favourable to the ARs deployment have allowed to continue to experiment and improve scientific knowledge on the contribution of ARs to the socio-ecosystem. The two study projects have emerged in this context.

3.2. The English Channel, a territory of research and development

The origin of the Etretat ARs project, located in the eastern part of the English Channel, is multiple and responds to various actor's interests (Figure 2):

- **The industry sector** is interested in experimenting and designing new materials, resistant in the corrosive marine environment.
- **The harbour manager** considers ARs project as an opportunity to increase harbour activity and enhance the harbour value through innovative projects.
- **Professional fishermen organizations** are attracted by the possibility of increasing the coastal fishery resource. However, the location of the project is essential and must not impact existing activities. The choice of the project area can then influence their support for the project.
- **Other marine users** such as divers are interested in developing more diving sites.
- **Marine fauna and flora** are interested in the possibility of having a specific and additional habitat to feed and reproduce.

Thus, all these actors and their interests converged around a common project of experimentation of form and materials of ARs, making this project the 1st in the English Channel.

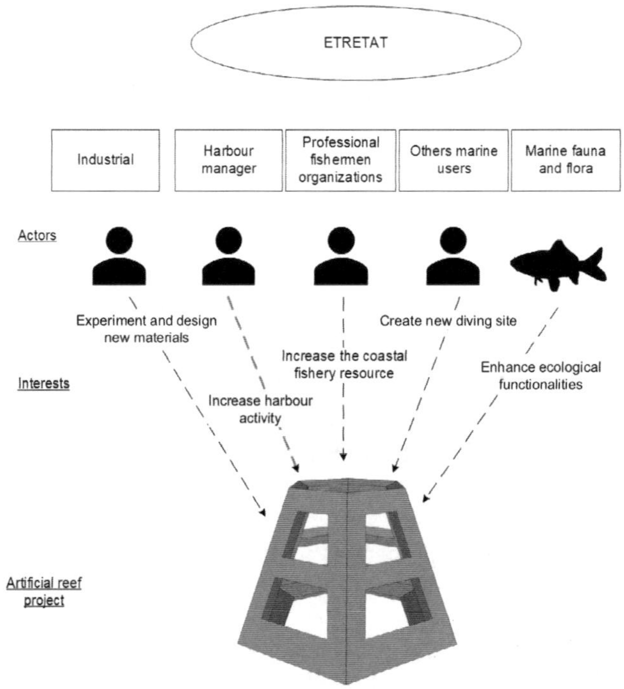

Figure 2. Actors interests in the context of the Etretat study site

The deployment of ARs in the Cherbourg harbour is an example of a convergence of several interests which can be divided into two main steps.

Firstly, the European INTERREG program entitled RECIF which has been able to federate scientific and industrial actors around a project to recover shellfish waste in the manufacture of concrete (Figure 3) :

- **The industry sector** continues their research and development approach. The evolution of the regulatory context (compensatory measure), economic context (offshore wind turbine project) and mentalities in favour of ecology, suggests that ARs project could be promising subject.
- **The scientific community** is also interested by study human pressures on the environment, both in terms of increasing industrial waste such as shells and the increase in activities in the marine environment. Through a research program, researchers are interested in developing new materials, monitoring protocols and acquiring data on the flora and fauna of hard substrates that had previously been neglected in terms of their biodiversity and ecological functioning at the scale of the French side of the English Channel.
- **Marine fauna and flora** are interested in the possibility of having a specific and additional habitat to feed and reproduce.

Secondly, a consultation around the Cherbourg harbour development project invites users to give their opinion and express their interests. The deployment of ARs appears as an interesting ecological action to compensate the harbour extension which can have a negative effect of the coastal biodiversity. The idea then reached a consensus among users of the Cherbourg harbour.

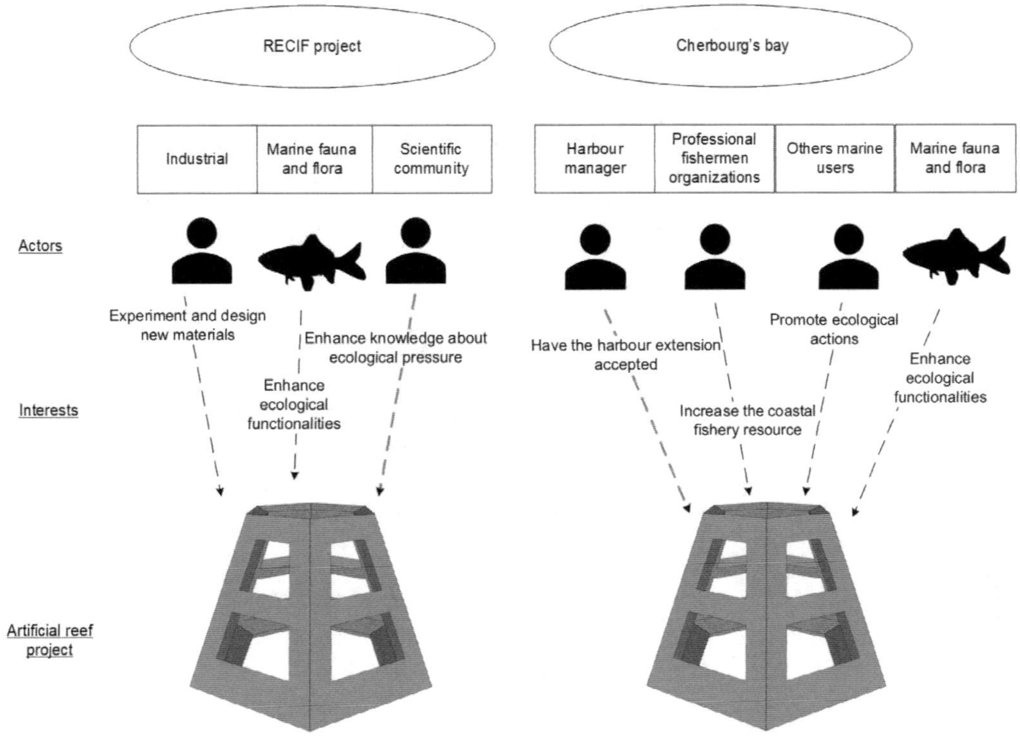

Figure 3. Actors interests in the context of the Cherbourg study site

3.3. Actors enrol in AR's immersion project

In order to carry out ARs projects, the previously federated actors are involved in their deployment and also mobilize other actors necessary for the project. The needs for the deployment of ARs are numerous. First of all, there is an administrative need for authorization to use the maritime French public domain, then a need for financing the project and finally a need to define the technical aspects of design, deployment and monitoring of the structures.

For the Etretat project, the actors involved will fill the roles of project holder (legal project leader for the administration), project manager, operator and marine user (concerted for the define the location and the management measures). The interests of marine fauna are considered at this stage of the project, through scientific monitoring and a preliminary study to determine the best location of the ARs. For funding, the actors will request European funds and local authorities support. Regarding the need of authorization, it is given by the administrators of maritime and land affairs (DDTM).

The RECIF program and the reef project in the harbour of Cherbourg are not on the same temporalities. While the RECIF project is concretized from a technical and financial point of view, the Cherbourg harbour manager carries out a preliminary study in order to choose the ARs location. The needs and interests of these two projects will then converge to form a single ARs project. Indeed, while the RECIF project encounters administrative difficulties (duration of instruction exceeding the temporality of the European program), the project leader of the Cherbourg harbour has administrative facilities. The merger of the two ARs projects then brings together all the administrative, financial, technical and scientific needs to implement ARs (Figure 4). Finally, the monitoring of the ARs deployed in Cherbourg harbour continues today as part of a new European INTERREG MARINEFF project and bringing together the actors of the RECIF project as well as other partners including the Cherbourg harbour manager for the deployment of new reefs in the Cherbourg Bay.

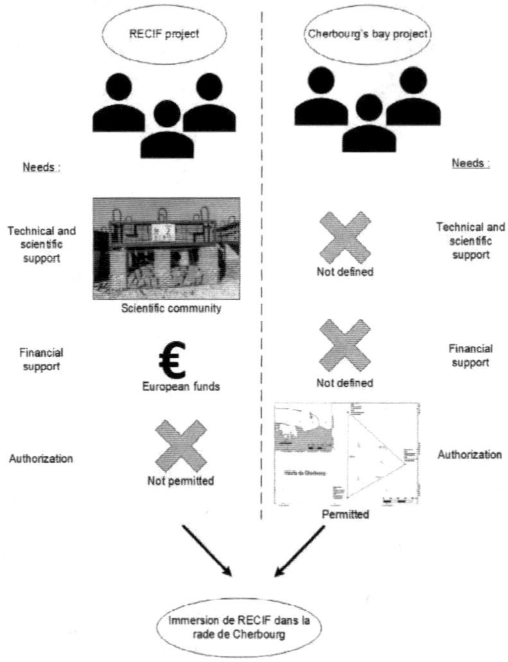

Figure 4. Mobilization of actors in the context of artificial reefs deployment in the harbour of Cherbourg

3.4. The continued mobilization of actors

A project is defined as a success when all actors enroll in the project to achieve the ARs deployment [16]. The continued mobilization of actors ensures the achievement of the results expected by each actor.

In the case of the Etretat ARs, the actors involved have different economic, social or ecological expectations. After ten years of deployment, the feedbacks from the actors involved in the project are:

- **The industry sector**: it has succeeded in promoting the experimentation of a formulation of new materials for which a patent has been filed.
- **The harbour manager:** despite a lack of interest of wind project leaders in artificial reefs, the harbour has been able to position as a logistics platform for the installation of wind farms, harbour activity has increased accordingly. However, the influence of the ARs project on this increase is difficult to determine.
- **Professional fishermen organizations** don't have specific benefit related to the ARs deployment and subsequently lost interest in the subject.
- **Others marine users** : divers benefit from a partnership with the harbour manager. They carry out annual photographic monitoring on the ARs.
- **Marine fauna and flora**: it colonized the AR with a preference for cubic reefs, compared to the large central module. Despite the destabilization of the central module, the diversity of the fauna increases.

In the case of the RECIF project in Cherbourg harbour, after five years of immersion, the feedback of deployment, the feedbacks from the actors involved in the project are:

- **The industry sector:** they continue their research and development approach in the new MARINEFF research program.
- **The harbour manager**: it was able to carry out the harbour extension and compensation. He joined the new MARINEFF research program as a partner.
- **The scientific community**: it has promoted research work through the realization of a scientific symposium. In addition, it is continuing to acquire biological data within a new INTERREG programme called MARINEFF.
- **Others marine users**: they accepted the ARs project. The area being forbidden to all activities, they lost interest in the subject. Only one diving structure has become a technical partner of the scientific community.
- **Marine fauna and flora**: The flora has strongly colonized the reefs and the fauna is also present.

4. Conclusion

Through the translation concept, the process of building ARs projects is described. Initially isolated actors with their own interests will federated in a common project of ARs deployment. The ARs appear as the tools that permit to meet the expectations of all the actors involved. This approach makes it possible to describe the social expectations of the actors involved in the projects but also to integrate the ecological aspects of the projects by integrating the marine fauna and flora as an actor in its own right. A network of actors is then created around ARs projects.

The difficulty of assess an ARs project lies in the complexity of the socio-ecosystem formed during its deployment. Indeed, the translation concept highlights the diversity of actors interests. These interests, if the project is settled, can be used as goals to reach in order to measure the success of AR projects. There is therefore not one clearly defined objective but a variety of expectations depending on the actors initially enrolled and mobilized later.

The limits of the assessment by this approach are that it is a qualitative method and that the ecological dimension appears as minor in the process. Indeed, the interest of the ecosystem is considered through a single group that represent the marine fauna as a whole that is a very simplified analysis.

To go further in the understanding and assessment of the socio-ecosystem, the analysis could propose a modelling of the sociological and ecological ARs network. These models would bring quantitative indicators to the assessment and restore symmetry between human and non-human actors. By comparing the number of interactions before and after ARs deployment, it would be possible to characterize the network evolution over the life time of ARs.

5. Acknowledgments

This research is co-funded by National Association for Technical Research and Atlantique Landes Récifs' association. We want to thank all interviewed actors for their time.

6. References

[1] Lacroix D, Charbonnel E, Dao J-C, Véron G, Lagardère J-P, Mellon C, Covès D et Buestel D 2002. Les récifs artificiels. Aménagement du littoral marin et repeuplement. D. Lacroix Edit., IFREMER Publ., 148 p.

[2] FAO 2015. Practical guidelines for the use of artificial reefs in the Mediterranean and the Black Sea, by Fabi G, Scarcella G, Spagnolo A, Bortone SA, Charbonnel E, Goutayer JJ, Haddad N, Lök A, Trommelen M, Rome, Italy. Studies and Reviews, General Fisheries Commission for the Mediterranean 96, 84 p

[3] Salaün J, Pioch S, Dauvin J-C 2020. Artificial reef along the French Mediterranean coastline: toward innovative integrated biodiversity management, Firenze University Press, Proceedings e report : 309-315. doi.org/10.36253%2F978-88-5518-147-1.31

[4] Baine M 2001. Artificial reefs: A review of their design, application, management and performance. *Ocean Coast Manage* **44**: 241-259.

[5] Véron G, Denis J, Gérard A, Thouard E et Thébaud O 2008. Les récifs artificiels, état des connaissances et recommandations, Ifremer, 25 p.
[6] Ramos J, Santos M, Whitmarsh D and Monteiro CC 2008. A socio-economic toolbox of artificial reef projects. Conference: 14th IIFET 2008, Vietnam Proceedings, 12 p.
[7] Tessier A, Bodilis P, Charbonnel E, Dalias N, Francour P, Lenfant P and Seaman W 2015. Assessment of French artificial reefs: due to limitations of research, trends may be misleading. *Hydrobiologia* **753**: 29 p.
[8] Claudet J 2006. Aires marines protégées et récifs artificiels: méthodes d'évaluation, protocoles expérimentaux et indicateurs. Perpignan. Thèse de l'Université de Perpignan, 255 p.
[9] Cresson P, Ruitton S, Ourgaud M and Harmelin-Vivien M 2014. Contrasting perception of fish trophic level from stomach content and stable isotope analyses: A Mediterranean artificial reef experience. *J. Exp. Mar. Biol. Ecol.* **452**: 54-62. doi.org/10.1016/j.jembe.2013.11.014
[10] Cresson P, Le Direach L, Rouanet É, Goberville E, Astruch P, Ourgaud M and Harmelin-Vivien M 2019. Functional traits unravel temporal changes in fish biomass production on artificial reefs. *Mar. Environ. Res.* **145**, 137-146. doi:10.1016/j.marenvres.2019.02.018
[11] Milon JW, Holland SM and Whitmarsh DJ 2000. Social and economic valuation methods. In: Seaman W (ed.) Artificial reef evaluation with application to natural marine habitats, Boca Raton, Florida. CRC Press.: 165-194.
[12] Lee MO, Otake S and Kim JK 2018. Transition of artificial reefs (ARs) research and its prospects. *Ocean Coast Manage* **154**: 55-65. doi.org/10.1016/j.ocecoaman.2018.01.010.
[13] Tessier A, 2013. Caractérisation des Récifs Artificiels du Golfe du Lion : de l'écologie aux usagers. Thèse de doctorat de l'université de Perpignan, 316 p.
[14] Castetz N 2008. A Etretat, un récif artificiel pour revigorer les fonds. Libération.
[15] Auzas M et Maurouard. Immersion des récifs. Rapport technique 1, 22 p.
[16] Callon M 1986. Éléments pour une sociologie de la traduction : la domestication des coquilles St-Jacques et des marins pêcheurs dans la baie de St. Brieuc. *Ann Sociol* **36**:169-208.
[17] Latour B 1987. Science in Action: How to Follow Scientists and Engineers through Society. Harvard University Press, Cambridge, MA, 288 p.
[18] Crozier M et Friedberg E 1977. L'acteur et le système. Ed., Seuil, 500 p.
[19] Reverdy T 2013. Sociologie des organisations. Master. Sociologie des organisations, Grenoble-INP, France. 160 p.
[20] Lombard-Latune J 2018. La compensation écologique : du principe de non perte nette de biodiversité à son opérationnalisation – analyse de l'action collective, Thèse de doctorat de l'Université Paris-Saclay, 246 p.
[21] Alami S, Desjeux D et Garabuau-Moussaoui I 2013. Les méthodes qualitatives, 2e édition. ed. Paris : Presses Universitaires De France, 2009. Print. Que Sais-je ? Société 259 p.
[22] Kaufmann JC 2016. L'entretien compréhensif, coll 128 tout le savoir, 4ème ed. Armand Colin 2011, 128 p.
[23] Meur-Ferec C 2006. De la dynamique naturelle à la gestion intégrée de l'espace littoral : un itinéraire de géographe. Géographie. Université de Nantes, 248 p.
[24] FAO, 1998. La situation mondiale des pêches et de l'aquaculture, SBN 92-5-204187-7, FAO Rome
[25] Duval C et Duclerc J 1985. Les récifs artificiels en Languedoc-Roussillon. Etude halieutique : méthodologie et premiers résultats. IFREMER, 140 p.

A Review of the Ecosystem Services Provided by the Native Oyster (*Ostrea edulis*): Implications for Restoration

S Thomas, K Collins*, C Hauton and A Jensen

Ocean and Earth Science, University of Southampton, SO14 3ZH, UK

*Ccorresponding author email: kjc@noc.soton.ac.uk

Abstract. Native oyster (*Ostrea edulis*) habitat has been decimated across the majority of its natural range as a result of human activity. In recent decades, oyster restoration projects have gathered increasing support due to their potential to provide ecosystem services that offset increasing pressure from human development, resource demand, and also climate change. These ecosystem services are reviewed here, so as to inform the potential benefit of restoration projects. Ecosystem services can be divided into four categories: *provisioning services* through direct utilisation of oysters and the species they support, *regulating services* that help maintain a healthy environment, *habitat services* that benefit biodiversity, and *cultural services* that can influence tourism and wider cultural values. Ecosystem services are often interlinked in complex networks, though these can be linked back to two overarching ecosystem functions, stemming from either the filter feeding activity of oysters, or the physical reef structure that they provide. The MARINEFF oyster enhancement reefs installed in the Solent, September 2020, are intended to support the extensive Solent Oyster Restoration Project initiated by the Blue Marine Foundation. The expectation is that they will facilitate both oyster settlement and the release of larvae for the repopulation of the Solent system. This may contribute to the recovery of ecosystem services mentioned in this review. It is important to note however that there is a lack of baseline data for native oysters, and the ecosystem service values they provide are poorly constrained, meaning that ecosystem service estimates are derived largely from other species of oyster. This inhibits our ability to accurately calculate and model natural capital provided by native oysters, highlighting the necessity for further research so as to inform management and restoration targets.

1. Introduction

Ecosystem services (ES) can be defined simply as "the contributions that ecosystems make to human well-being" [1], and can be provided via extraction, utilisation, or experience [2]. The stock of a given ES can be referred to as 'natural capital' (NC), providing a means of quantifying their value in a given area [3]. Coastal habitats make up 6% of global biomes, however they provide approximately 37.8% of global ES values each year [4]. Epifauna, which include oysters, are considered amongst the highest providers of ES in the coastal system due to their wide distribution and range of ecosystem functions [5]. In Europe and North Africa, native oysters, *Ostrea edulis*, can be found predominantly in estuaries and sheltered coastal seas, in depths of up to 80m on mixed hard substrates [6]. Historically, native oysters have provided a multitude of ES, however due to overfishing and development, their numbers are now desperately low [7]. There has been growing support to make ES a focal point of marine management strategies [5,8]. In a 25-year environmental management plan outlined by HM Government in 2018 [9], it is stated that:

"When we give the environment its due regard as a natural asset – indeed a key contributor – to the overall economy, we will be more likely to give it the value it deserves to protect and enhance it. This is why, as signalled in our Industrial Strategy, over coming years the UK intends to use a 'natural capital' approach as a tool to help us make key choices and long-term decisions."

The MARINEFF project, a collaboration between the UK and France and funded by Interreg, aims to contribute to the restoration of native oysters and the ES they provide through the implementation of artificial reef modules [10]. This paper presents a review of the ES provided by native oysters, so that the potential benefits of this project and other schemes alike can be more fully realised. The capacity of native oyster restoration to enhance ES provision is also considered, however there is a large deficit of research on native oyster (*Ostrea edulis*) ES. As such, for the purpose of this paper, information and data gaps are accounted for based on studies of comparable species: Eastern oyster (*Crassostrea*

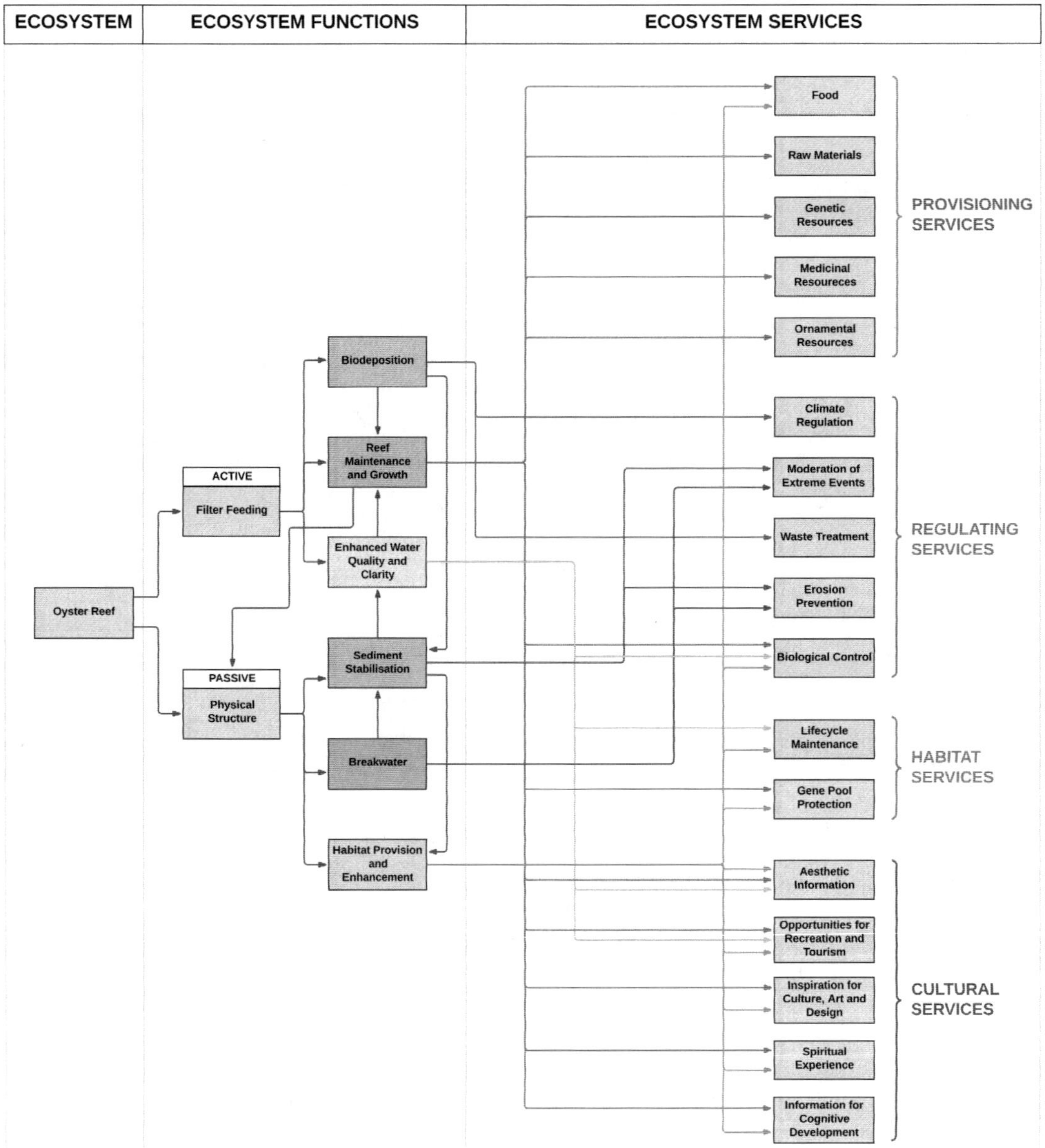

Figure 1. The network of ecosystem functions that underpin native oyster ecosystem services as identified within the TEEB typology [13]. A three-step process is used to indicate how an ecosystem produces a complex network of ecosystem functions, that can individually or collectively lead to ecosystem services valued by humans.

virginica), Pacific oyster (*Crassostrea gigas*), *Crassostrea rhizophora*, and blue mussel (*Mytilus edulis*).

2. Classifying Ecosystem Services

Native oysters provide ES either (i) passively, via the physical structure presented by oyster reefs, or (ii) actively, through their role as filter feeders [8]. The ES provided by these two functions are often coupled, creating a complex network of ecosystem functions that govern ES provision (Figure 1). ES can be classified in a number of ways, with many studies using or modifying existing classification frameworks, creating disparity in how ES are valued [3]. There is therefore support for the use a more standardised approach to ES assessments, maximising transferability of data [3,11]. A number of ES typologies exist, originating with the Millennium Ecosystem Assessment [12]. Following on from this, 'The Economics of Ecosystems and their Biodiversity (TEEB)' ES typology was created [13]. More recently, the 'Common International Classification of Ecosystem Services (CICES)' ES typology has been presented as an alternative to TEEB, and is now used widely within the EU [1,3]. However, this typology was created primarily for the terrestrial system, presenting challenges when applied to the marine environment [5]. For the purpose of this review, the TEEB typology is utilised for assessment of native oyster ES. This updated typology includes 22 ES divided into four categories: Provisioning Services, Regulating Services, Habitat Services, and Cultural Services [13,14]. Out of the 22 potential ES, 17 apply to native oysters (Figure 1). ES are most commonly valued in economic terms, allowing values to be incorporated into habitat management (Table 1).

Table 1. Economic value estimates of ecosystem services provided by oyster reefs. Estimates are missing for a number of ecosystem services, as many services are not directly quantifiable.

TEEB Classification	Ecosystem Service	Species	Economic Value	Reference
Provisioning Services				
Food	Commercial Value	Native Oyster	$13 per kg	FAO, 2021 (15)
	Commercial Fish Enhancement	Eastern Oyster	$4,123 per ha	Grabowski et al, 2012 (16)
Regulating Services				
Climate Regulation	Carbon Removal	Eastern Oyster	£28.80 per ha per yr	Watson et al, 2020 (17)
Moderation of Extreme Events	Breakwater	Eastern Oyster	$0 - $85,998 per ha per yr	Grabowski et al, 2012 (16)
Waste Treatment	Nitrogen Removal	Eastern Oyster	£12,774 per ha per yr	Watson et al, 2020 (17)
	Phosphorus Removal	Eastern Oyster	£2,483 per ha per yr	Watson et al, 2020 (17)
Habitat Services				
Lifecycle Maintenance	Submerged Aquatic Vegetation Enhancement	Eastern Oyster	$1,292 per ha per yr	Grabowski et al, 2012 (16)

2.1. Provisioning Services

2.1.1. Food. Harvest of native oysters provides a direct food source, whilst oyster reef habitat enhances marine biodiversity, supporting the ecosystems that underpin commercially important species of fish (6). A study by Peterson et al [18] found that $10m^2$ of restored eastern oyster reef produces approximately 2.6 kg of fish and large crustaceans per year. Grabowski et al [16] reported that a single hectare of eastern oyster reef produces approximately $4,123 of commercial fish via direct harvest per year. It is worth noting however that the enhancement of non-oyster fisheries is highly spatially variable, meaning location specific information is required when making estimates [19].

2.1.2. Raw Materials. The value of harvested oysters is derived almost entirely from the food source they provide. However, the edible component represents less than 10% of oyster mass, with the rest being shell material [20]. Due to the large amounts of oysters that are produced globally for the food industry, oyster shell waste has led to environmental issues relating to improper disposal [21]. However, oyster shell presents a renewable by-product of oyster harvest that can provide added commercial value to the harvesting of oysters [20]. In most cases, powdered oyster shell (POS) provides the most widespread potential for industrial application [20].

In construction, concrete produced with 10% POS has been observed to present the same properties as normal concrete, as well as having a shorter hardening time and a higher initial compressive strength [22,23]. In agriculture, soil acidification has become a prominent issue as a result of excessive use of nitrogen based fertilisers, which if not taken up by plants are converted to nitrate in the soil, releasing hydrogen in the process and therefore decreasing soil pH [24]. Oyster shell is largely comprised of calcium carbonate, which when introduced to soil increases alkalinity, providing a buffer to soil acidity [25]. POS can also be fed directly to farmed animals, and is frequently used as supplementary calcium carbonate for the diet of farmed chickens to improve eggshell quality [26]. Oyster shell possesses potential for waste-to-energy schemes through the production of biofuel. A study by Nakatani et al [27] found that by combusting oyster shell at 700°C, the product can be used as a catalyst for the reaction that converts soybean oil and methanol to a viable form of biodiesel. In a similar study, Choi et al [28] found that pyrolysis of brown algae at 720°C could be effectively catalysed by the calcium carbonate contained within oyster shell, producing viable synthetic natural gas.

2.1.3. Genetic Resources. A study on *Crassostrea Rhizophora* in Brazil found that nacre, the inner shell material of oysters, presents suitable properties for application as a potential bone substitute [29]. The experiment, conducted on rats, found that oyster nacre provides a biocompatible material that promotes bone formation successfully, highlighting its potential osteological applications for humans.

2.1.4. Medicinal Resources. Oyster shell has a wide range of potential applications for human health. POS has abrasive properties, and as such is used in skincare products as an exfoliant, presenting a more sustainable alternative to plastic microbeads [20]. Oyster shell also provides a source of calcium; laboratory studies on rats have indicated that a diet fortified with POS can reduce the risk of osteoporosis and stimulate bone recovery from osteoporosis [30,31]. The use of oyster shell for human consumption and skin application requires further research, however. Oyster shell can bioaccumulate heavy metals, meaning that potential implications of consumption upon human health must be further assessed [20,32].

2.1.5. Ornamental Resources. Whilst native oysters do not produce pearls or nacre of commercial value, its shell has been used for jewellery, art, as well as for decorative purposes [26].

2.2. Regulating Services

2.2.1. Climate Regulation. The role of oysters in carbon removal is not fully understood, however it is evident that oysters play a significant role in carbon cycling through benthopelagic coupling, the linkage of biogeochemical processes that occur in seabed sediments and the overlying water column [6]. There are no complete estimates of carbon removal potential for native oysters, though Watson et al [17] compiled multiple estimates of carbon removal in eastern oysters, and indicated that they can be considered a net sink or source of carbon, depending on whether the mean (-105 kg per hectare per year) or median (40 kg per hectare per year) value is used (Table 2). They suggested that the large variability in data warrants using the median value, and using this value calculated that based on UK CO_2 abatement cost of £60 per tonne of CO_2, carbon removal conducted by oysters is a service worth £28.80 per hectare per year. This is low value service, but it is important to note that this estimate is based solely upon abatement costs, meaning that the longer-term value of carbon sequestration as part of climate change mitigation is not considered. Moreover, there is a requirement for further research specific to carbon removal in native oysters.

Lee et al [33] assessed carbon deposition in native oysters occurring through both active deposition of pseudofaeces, as well as passive deposition of particulate matter through the altering of hydrodynamic processes that oyster reefs induce. The lab-based study found that in population densities of 75 oysters per m^2, native oysters deposited 2.62g of organic matter per day, with organic carbon comprising 0.14g of this amount. They highlighted that other biogeochemical processes would need quantifying to understand the full potential for native oysters to sequester carbon however, as carbon deposited can be regenerated, assimilated, or sequestered. Furthermore, fluxes of respiration and calcification would also need quantifying to establish the net effect of oysters as carbon sources or sinks. Respiration releases carbon, whilst calcification, despite removing carbon through shell assimilation, has a net effect as a carbon source [34]. This is because the calcification reaction results in the formation of carbonic acid, which dissociates to CO_2 and can then vented back into the atmosphere [34]. The net effect of all of these processes can be explained by the following equation [33]:

Net deposition = (Biodeposition + Passive Sedimentation + Calcification) – (Respiration + Calcification)

Table 2. Mean and median rates of nitrogen, phosphorus, and carbon removal for eastern oyster (adapted from Preston *et al.* [38]).

	Nitrogen (kg ha yr)		**Phosphorus (kg ha yr)**		**Carbon (kg ha yr)**	
	Mean	Median	Mean	Median	Mean	Median
Eastern Oyster	185	43	23	7	-105	40

2.2.2. Moderation of Extreme Events. Oyster reefs act as breakwaters, reducing wave energy and thus flood potential from storm surges [35]. Grabowski *et al* [16] estimated that the value of flood protection provided by eastern oyster is as much as $85,998 per hectare per year in areas of high levels of human development. Hard coastal protection structures have historically been the option of choice for flood defence when compared to restoring natural flood defence structures such as oyster reefs [36]. This is largely because they are cheaper to install, are immediately effective, and provide a more predictable degree of coastal protection [37]. However, hard structures can increase bottom erosion and alter sediment supply, which can result in habitat destruction [36]. They are also prone to degradation over time, requiring costly maintenance work [36].

Whilst oyster reefs provide a less quantifiable and predictable degree of coastal protection, they are self-sustaining and repairing, and thus do not incur maintenance costs unless they are subject to disease or destruction from human activity [37]. Moreover, oyster reefs grow vertically, at a rate faster than predicted sea level rise associated with climate change [16]. They may therefore be able to migrate spatially or accrete vertically in response to sea level rise, providing a degree of adaptability to climate change that hard structures do not [36]. Given the benefits and drawbacks of both natural and hard coastal defence structures, it is apparent that by combining both natural and hard engineering, integrated coastal zone management schemes could provide the most effective, economic, and environmentally sustainable approach to coastal management [37,38].

2.2.3. Waste Treatment. Oysters graze on phytoplanktonic algae, and in doing so remove seston; organic and inorganic material suspended within the water column [39]. A single oyster can filter as much as 200 litres of water per day, meaning they are important regulators of water quality [6]. Oyster filter feeding also provides a top down control that reduces phytoplankton biomass and likelihood of harmful algal bloom emergence [40]. Nutrients contained within seston can be passively deposited via the physical oyster reef structure, assimilated into oyster tissue and shell, or concentrated into faeces or pseudofaeces, collectively referred to as biodeposits. Nutrients entering the underlying sediment are subject to a range of biologically facilitated cycling processes, with a large amount either recycled within the ecosystem or regenerated back into the water column [41]. Cycling of nutrients within oyster reef habitat does not operate at 100% efficiency however, meaning that nutrients that are not recycled or regenerated can be permanently removed from the system. This occurs either via sequestration within the sediment, or via bacterially facilitated conversion of bioavailable nitrogen to biounavailable nitrogen through denitrification and anammox processes [41]. Commercial oyster harvest also represents a nutrient sink, whereby nutrients assimilated in oyster tissue and shell are permanently removed from the marine system [42,43].

Based on a number of studies, Watson *et al* [17] estimated median nitrogen (N) and phosphorus (P) removal for eastern oyster at 43 and 7 kg per hectare per year respectively (Table 2). Based on UK wastewater treatment costs, they calculated that N and P removal by oysters is a service worth £12,774 and £2,483 per hectare per year, respectively. The waste removal cost per kg of nutrients increases as nutrient concentrations decrease, with nitrogen removal being the more costly process [17,26]. This means that enhanced nutrient removal through restoration could provide an important service in lower nutrient conditions, saving water treatment costs [26]. Caution is to be taken when comparing removal estimates between species of oyster however; eastern oysters are larger, form denser beds, filter greater volumes of water and therefore likely have higher removal rates than native oysters, meaning that rates of nutrient removal could be significantly different between species [44].

2.2.4. Erosion Prevention. The breakwater function of oyster reefs can reduce coastal erosion, protecting the human environment as well as surrounding natural habitats [35]. Oysters have been observed to be capable of attenuating wave height by as much as 95%, though this is dependent upon a number of factors including reef depth, height, width and length [38]. One study found that a 20 metre wide oyster reef can reduce wave energy reaching the coastline by as much as 70% [45]. Another study found that breakwaters made from oyster shell were capable of reducing coastal erosion and retreat by as much as 40% [37].

Research has indicated that the ecosystem engineering capabilities of an oyster reef can influence sediment dynamics of area beyond the reef itself, to the same order of magnitude as the area of the reef. Walles *et al* [46] observed that in Pacific oyster reefs, an accumulation of sediment created an elevated area on the lee of the reef that reached 40% of the height of the reef itself. They found that the size of the area of influence was dependent mainly on reef length, though it was also noted that reef height displayed the highest potential for wave energy dissipation, highlighting the potential of vertically accreted reefs as breakwaters. As a result, it was suggested that oyster reefs could provide the highest level of protection from coastal erosion when placed seawards of the area requiring protection. This is because the area of influence would be much larger when placed seaward, promoting sedimentation, and reducing erosion.

2.2.5. Biological Control. Restored and healthy reefs are self-sustaining; there is evidence to suggest that vertical complexity in pristine reefs provides resilience to disease. One study found that oysters at the top of vertically accumulating reefs were subject to higher flows, higher food quality, and lower sedimentation, bringing about low infection rates from diseases [47].

2.3. Habitat Services

2.3.1. Life Cycle Maintenance. The term 'habitats for species' describes the value that an individual species derives from the habitat that supports it, regardless of human benefit. Of course, this can also lead to human benefit, but the value here is intrinsic [13]. This provides a non-anthropocentric service without means of economic valuation, but one that must be recognised for its importance nonetheless [48]. Oysters are reef builders, creating habitat complexity that provides refugia, nesting habitat, and also nursery grounds for juvenile fish. They also support rich algal and microphytobenthic communities, providing food for grazers that live within the reef [6]. By better understanding the ecological interactions that occur between species and their spatial variation, we can better appreciate and value them for their contribution to nature alone [49]. Studies have shown that bivalves, in particular oysters, are capable of increasing species abundance and richness, as well as overall biomass [35,50]. Oyster reefs are structurally complex, providing an abundance of habitat to support marine life [8]. As previously mentioned, oyster reefs can influence hydrodynamics in an area that reaches far beyond the oyster reef itself [46]. In a similar study on mussel beds in the Wadden Sea, Donadi *et al* [51] corroborated this idea. They found that the far-reaching alterations in sedimentary regime induced by mussel bed bioengineering resulted in increased habitat suitability for communities of cockle to reside at around 50 to 100 metres away from the mussel bed itself. Understanding the potential for oyster reefs to influence habitat and species distribution around them could therefore prove an important asset for efficient and effective integrated habitat restoration programmes.

The biodeposits that oysters produce can enrichment sediments, stimulating growth of submerged aquatic vegetation (SAV) through a 'spill over' effect [52,53]. Furthermore, the improvements in water clarity that oysters provide can benefit submerged aquatic vegetation (SAV) directly by increasing photosynthetically active radiation (PAR) in the benthos [35]. Grabowski *et al* [16] estimated that SAV enhancement from one hectare of eastern oyster reef is worth approximately $1292 per year. The breakwater function of oyster reefs can also promote sediment stabilisation and deposition, enhancing SAV growth and protecting adjacent SAV habitats such as saltmarshes and seagrass from erosion and degradation [6].

2.3.2. Gene Pool Protection. Research has shown that the evolutionary success of one species can be important to that of another, introducing the concept of 'survival value' [48]. This raises the idea that by simply conserving the species that we interact with, we promote the complex relationships between species that facilitate evolutionary success. This is a means of giving value to survival itself, as if it were the purpose of any given ecosystem regardless of any conscious endeavour. In this context, keystone species, which include oysters due to their bioengineering potential, possess a particularly high survival value [54]. The vast decline in native oyster populations by up to 95% in Europe has depleted this survival value, posing negative implications for the survival of species that benefit from its success [6]. Moreover, this decline in population has decreased the size of the native oyster gene pool, meaning that restoration efforts will be working to restore native oyster populations with a lower genetic diversity. This poses risks relating to species resilience, whereby the potential impacts of natural events and disease outbreaks could be intensified [55].

2.4. Cultural Services

2.4.1. Aesthetic Information. The biodiversity that oyster reefs support provides a natural environment valued by those who enjoy snorkelling, diving, and birdwatching [26]. Improved water quality and clarity promoted by oyster filter feeding can lead to cleaner visibly cleaner water, creating an aesthetic of a cleaner and healthier environment [56]. Furthermore, in the context of coastal defence, oyster reefs provide more aesthetic and less restrictive environment, which can in turn benefit the local economy [38].

2.4.2. Opportunities for Recreation and Tourism. Enhanced water quality and clarity provided by the filter feeding activity of native oysters can improve recreational experience. Holidaymakers and regular recreational users are encouraged to use waters of high bathing quality, boosting the local economy [8]. The biodiversity supported by native oyster habitat can provide opportunity for recreational fishermen and nature enthusiasts, attracting visitors [26]. The ecotourism industry also depends heavily on high water quality and biodiversity. For example, oyster reefs provide a food source and habitat for fish, facilitating a transfer of energy through trophic levels that ultimately supports larger animals that provide economic opportunity through ecotourism, such as dolphin populations [57]. There is also value derived from the symbolic presence of habitats such as oyster reefs as places of natural beauty. Pristine ecosystems can provide meaningful places and socially valued spaces, promoting tourism through those who come to enjoy the environment they create and observe the species that they support [57]. The consumption of oysters can also promote food tourism through regional food traditions, as well as for food festivals [26].

2.4.3. Inspiration for Culture, Art and Design. Oysters have inspired culture in all forms, from the arts, to fashion, to architecture. For example, the electroacoustic signals produced by eastern oyster reefs can be monitored using bioacoustics and have been utilised in the musical arts to create a novel natural soundscape. This reflects how the beauty of the marine environment can be expressed through different mediums, allowing people to engage with and appreciate its complexity in unique ways [58]. Native oyster shell has also been used to make jewellery items, as well as for decorative purposes on buildings [26]. Furthermore, oyster shell is of archaeological significance, as evidence from midden piles have indicated that oyster shell was historically used to make tools [26].

2.4.4. Spiritual Experience. Natural habitats present value derived from bequest. In this case, the preservation of native oyster habitat provides a legacy for future generations, so that they may continue to enjoy the rich history associated with native oysters [26]. This extends into spirituality, with many deriving transcendences from the natural environment, giving a sense that without these habitats our world is less diverse and beautiful [55]. By conserving native oysters, we are preserving that value, whilst taking solace from the knowledge that this habitat exists and is healthy and thriving [26].

2.4.5. Information for Cognitive Development. There has been an increased support for marine science and conservation to be included as part of a school curriculum. In New York, the Billion Oyster Project [59] has collaborated with 100 schools, making a strong step towards providing educational resources, lessons, as well as field excursion opportunities relating to oyster restoration. Moreover, the project has operated in an inclusive manner, placing emphasis on those in minority communities and encouraging them to pursue a career in a STEM field.

3. Restoring Natural Capital

3.1. Feasibility of Restoration

Native oyster populations have declined significantly across their biogeographic range since the late 1800's, and so knowledge of baseline habitat conditions is poor as their decline occurred before records began [7]. In addition, there is a lack of ES values for native oyster habitat in the literature, meaning ES value estimates are largely taken from other species. This reduces the accuracy of NC enhancement calculations for restoration projects. Therefore, a greater understanding of native oyster ES values will facilitate better predictions of NC enhancement potential, subsequently improving marine management strategies and policy making in the absence of baseline data [60].

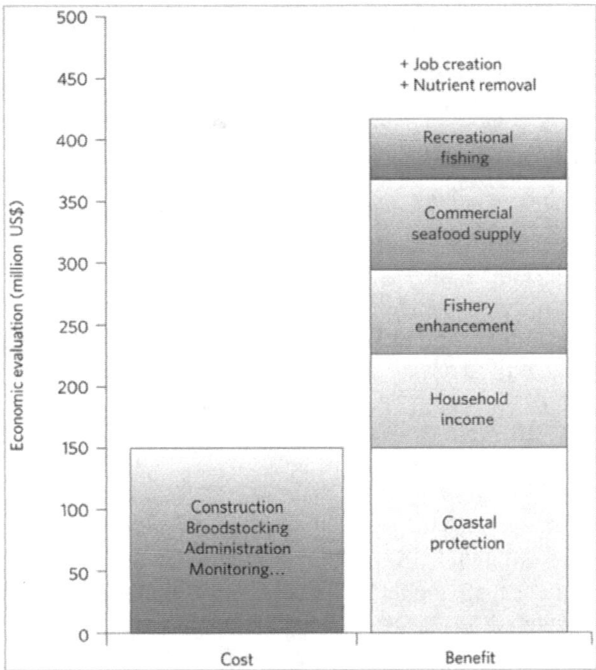

Figure 2. Composite bar chart indicating the comparative cost of oyster restoration (red) along with the NC benefits (green) associated with restoration of a 100-mile-long reef over a ten-year period, from Cheong *et al* [38].

Studies have indicated restoration as an economically viable action for improving NC values associated with oyster reefs (Figure 2) [38,61]. Grabowski *et al* [16] estimated that eastern oysters reefs recover the cost of initial restoration within 2 to 14 years through the NC that they provide. Moreover, these estimates are conservative, as they do not cover the full range of ES that contribute further to NC stocks. However, despite the recognised cost effective potential of oyster restoration, uncertainty in NC estimates can lead to a lack of action, whereby artificial alternatives are favoured as they provide a more immediate, quantifiable, and spatially and temporally consistent service [37].

Although restoration is economically viable, there are a number of factors that may complicate native oyster restoration efforts. The long standing depletion of native oyster habitat raises doubts about the feasibility of restoring a habitat that has existed in an altered state for over a century [8]. Many areas of historic oyster reefs have undergone a phase shift from benthic dominated to planktonic dominated communities, making restoration difficult as the ecosystem is now in a new state of equilibrium [16]. Paradoxically, native oysters preferentially settle on existing shell material, yet depleted stocks present little shell, inhibiting natural recovery and highlighting the requirement for active restorative measures [62]. Furthermore, continued overfishing, low natural settlement rate, and competition from non-native species also inhibit natural recovery [62]. For example, Native oyster populations face competition from the non-native slipper limpet, *Crepidula fornicata*, which can grow in high densities, outcompeting native oysters for space and producing deposits that alter seabed characteristics, preventing settlement and further inhibiting native oyster settlement [44]. Native oyster restoration is also complicated by diseases caused by parasitic protists such as *Bonamia ostreae* and *Marteilia refringens*, the former of which being a more serious problem that has caused population mortalities of up to 90% since its emergence in the 1980's [63,64]. Studies have indicated that higher native oyster densities can lead to increases in the spread of *B. ostreae*, presenting challenges when setting restoration targets [65].

3.2. Artificial Reef Restoration

Artificial structures are a popular choice for restoration of reef building habitats, as they provide suitable habitat in areas where it has been lost, creating a strong foundation for initial reef development [66]. Artificial structures also discourage bottom-trawling activities, meaning areas with structures present have the functional effect of a marine protected area, promoting conservation and allowing recovery of ES [57]. The MARINEFF project have designed artificial reef modules comprised of steel reinforced concrete, which are lined internally with bivalve shell material to facilitate oyster spat settlement [10, 67]. If the structures are successful in promoting native oyster recruitment, their design could be refined and implemented on a larger scale, restoring native oyster populations and the ES that they provide.

3.3. Case Study: The Solent Oyster Restoration Project

The Solent, a region of coastline in south England, has seen declines in native oyster populations since the late 1800's as a result of increased water pollution, habitat loss, overfishing, competition from invasive species, and disease [68]. In 2015, the Solent Oyster Restoration Project was set up by the Blue Marine Foundation, a collaborative operation that aims to introduce ten million native oysters to the Solent in an effort to re-establish a self-sustaining population of native oysters that support a viable fishery and provide other ES to the area [68]. The project, which is estimated to cost £1,436,500, aims to achieve this within 15-20 years, and integrates a number of restoration methods to maximise the project chances of success [68]. This includes establishment of sites prohibited from towed fishing gear, whereby broodstock oyster cages designed to promote the release of larvae will enhance spat settlement on the seabed and in spatting ponds for later relaying [68]. This method of oyster culture, in combination with sourcing oysters from hatcheries, increases gene pool variety, providing more opportunity for the emergence of individuals that present resistance and thus population resilience to diseases such as bonamiosis [68,69]. Natural spat settlement can then be enhanced using cultured or hatchery reared oysters, which are introduced directly to the seabed in sites identified as having the right hydrodynamic and substrate conditions for further spat settlement [68].

The oyster modules produced by the MARINEFF project can be integrated into the Solent Oyster Restoration Project. Currently, 26 modules of $1m^2$ each have been deployed within $100m^2$ of seabed in a pilot site at Calshot in the Solent [10]. Each module presents a surface area of $5m^2$, creating $130m^2$ of viable artificial reef surface, occupying $26m^2$ of seabed across all modules. The potential role of these modules within the Solent Oyster Restoration Project lies within their function as a passive means of restoration, providing substrate for settlement of oyster spat. Oyster modules could be placed in regions of seabed laying of mature oysters and spat-on-shell, increasing the available substrate for further settlement, and facilitating vertical complexity of restored reefs. Moreover, another potential usage of oyster modules could be via placement in areas of the Solent that have known transport of oyster larvae, but seabed substrate unsuitable for direct oyster settlement. These regions may have historically supported native oyster populations, but habitat loss may have allowed alteration of seabed sediment dynamics, inhibiting natural recovery. The oyster modules could help to initiate recolonisation in these areas, providing a suitable surface for settlement above the seabed. In turn, the bioengineering capabilities of native oysters could alter seabed characteristics via biodeposition and sediment stabilisation, potentially leading to restoration of viable oyster habitat and an increase in ES provision.

4. Conclusions

This review highlights the diverse range of ES that native oyster habitat provides through direct provision of goods and resources, regulation and maintenance of the marine environment, cultural significance, as well as ES that do not have a human component, and are instead intrinsic, promoting the preservation and survival of life. Native oysters provide crucial ES through their bioengineering and reef building capabilities, which provides habitat and increases biodiversity, as well as reducing flood risk and coastal erosion potential. Furthermore, benthopelagic coupling facilitated by oyster filter feeding reduces harmful pollutants and excess nutrients, improving water quality and clarity as a result. This in turn provides favourable conditions for other habitats to grow and thrive, whilst also providing better conditions for recreation and tourism. Oysters themselves are a valuable fishery, and by enhancing biodiversity they can support other commercial fisheries.

Native oyster habitat restoration is an economically viable means of restoring NC stocks, especially given that oyster ES provision is self-sustaining, providing long term economic benefit far beyond restoration costs, reducing the requirement for wastewater treatment plants and coastal defence structures. ES values for native oysters are poorly constrained in existing literature however, highlighting a requirement for research to provide these values, so that NC stocks can be more accurately calculated and the value of native oyster habitat fully realised.

Acknowledgements
This study was supported by the MARINEFF project, selected by the European cross-border Interreg IVA France (Channel)-England programme co-funded by the ERDF. Simon Thomas was employed through a MARINEFF funded internship.

References
[1] Haines-Young R and Potschin M 2018 *Common International Classification of Ecosystem Services (CICES) V5.1 and Guidance on the Application of the Revised Structure*
[2] Rullens V, Lohrer A M, Townsend M and Pilditch C A 2019 Ecological mechanisms underpinning ecosystem service bundles in marine environments - A case study for shellfish *Front Mar Sci* **6** 1–16
[3] Hooper T, Börger T, Langmead O, Marcone O, Rees S E, Rendon O, Beaumont N, Attrill M J and Austen M 2019 Applying the natural capital approach to decision making for the marine environment *Ecosyst Serv* **38** 1–12
[4] Costanza R *et al* 1997 The value of the world's ecosystem services and natural capital *Nat* **387** 253–60
[5] Culhane F E, Frid C L J, Royo Gelabert E, White L and Robinson L A 2018 Linking marine ecosystems with the services they supply: what are the relevant service providing units? *Ecol Appl* **28** 1740–51
[6] Preston J, Gamble C, Debney A, Helmer L, Hancock B and zu Ermgassen P 2020 *European Native Oyster Habitat Restoration Handbook*, ed J Preston *et al* (London: The Zoological Society of London)
[7] Fariñas-Franco J M, Pearce B, Mair J M, Harries D B, MacPherson R C, Porter J S, Reimer P J and Sanderson WG 2018 Missing native oyster (Ostrea edulis L.) beds in a European Marine Protected Area: Should there be widespread restorative management? *Biol Conserv* **221** 293–311
[8] zu Ermgassen P, Hancock B, DeAngelis B, Greene J, Schuster E, Spalding M and Brumbaugh R 2016 *Setting objectives for oyster habitat restoration using ecosystem services: A manager's guide* (Arlington VA: The Nature Conservancy)
[9] Her Majesty's Government 2018 *A Green Future: Our 25 Year Plan to Improve the Environment.* (London)
[10] Collins K J, Hauton C, Jensen A C and Mallinson JJ. 2021 Solent Oyster Enhancement Reef Concrete Colonisation Trials, MARINEFF Conference, Caen, 14-16 Sept 2021
[11] Potschin M and Haines-Young R 2016 *Routledge Handbook of Ecosystem Services*, ed M Potschin *et al* (London and New York: Routledge) pp 25–44

[12] Millennium Ecosystem Assessment 2005 Ecosystems and Human Well-being: Wetlands and Water (Online)

[13] TEEB 2010 The Economics of Ecosystems & Biodiversity: Mainstreaming the Economics of Nature: A synthesis of the approach, conclusions and recommendations of TEEB (Online) Available from: http://teebweb.org/publications/teeb-for/synthesis/

[14] De Groot R *et al* 2010 *The Economics of Ecosystems and Biodiversity: Ecological and Economic Foundations* (Netherlands: Wageningen University & Research) pp.9–40

[15] FAO 2021 Ostrea edulis (Online) Available from: http://www.fao.org/fishery/culturedspecies/Ostrea_edulis/en

[16] Grabowski J H, Brumbaugh R D, Conrad R F, Keeler A G, Opaluch J J, Peterson C H, Piehler M F, Powers S P and Smyth A R 2012 Economic valuation of ecosystem services provided by oyster reefs *BioScience* **62** 900–9

[17] Watson S C L, Preston J, Beaumont N J and Watson G J 2020 Assessing the natural capital value of water quality and climate regulation in temperate marine systems using a EUNIS biotope classification approach *Sci Total Environ* **744** 1–11

[18] Peterson C H, Grabowski J H and Powers S P 2003 Estimated enhancement of fish production resulting from restoring oyster reef habitat: Quantitative valuation *Mar Ecol Prog Ser* **264** 249–64

[19] Gregalis K C, Johnson M W and Powers S P 2009 Restored Oyster Reef Location and Design Affect Responses of Resident and Transient Fish, Crab, and Shellfish Species in Mobile Bay, Alabama *Trans Am Fish Soc* **138** 314–27

[20] Bonnard M, Boury B and Parrot I 2019 Key Insights, Tools, and Future Prospects on Oyster Shell End-of-Life: A Critical Analysis of Sustainable Solutions *Environ Sci Technol* **54** 26–38

[21] Silva T H, Mesquita-Guimarães J, Henriques B, Silva F S and Fredel M C 2019 The potential use of oyster shell waste in new value-added by-product *Resources* **8** 1–15

[22] Yang E I, Kim M Y, Park H G and Yi S T 2010 Effect of partial replacement of sand with dry oyster shell on the long-term performance of concrete *Constr Build Mater* **24** 758–65

[23] Hong M, Jang I, Son Y, Yi C and Park W 2021 Agricultural by-products and oyster shell as alternative nutrient sources for microbial sealing of early age cracks in mortar *AMB Expr* **11** 1–15

[24] Kwon Y T, Lee C W and Yun J H 2009 Development of vermicast from sludge and powdered oyster shell *J Clean Prod* **17** 708–11

[25] Lee C H, Lee D K, Ali M A and Kim P J 2008 Effects of oyster shell on soil chemical and biological properties and cabbage productivity as a liming materials *Waste Manag* **28** 2702–8

[26] Van der Schatte Olivier A, Jones L, Vay L Le, Christie M, Wilson J and Malham S K 2020 A global review of the ecosystem services provided by bivalve aquaculture *Rev Aquac* **12** 3–25

[27] Nakatani N, Takamori H, Takeda K and Sakugawa H. 2009 Transesterification of soybean oil using combusted oyster shell waste as a catalyst *Bioresour Technol* **100** 1510–3

[28] Choi D, Nam I H, Park Y K, Ok Y S, Lee J and Kwon E E 2019 Catalytic pyrolysis of brown algae using carbon dioxide and oyster shell *J CO2 Util* **34** 668–75

[29] Coringa R, de Sousa E M, Botelho J N, Diniz R S, de Sá J C, Nogueira da Cruz M C F, Paschoal M A B and Gonçalves L M 2018 Bone substitute made from a Brazilian oyster shell functions as a fast stimulator for bone-forming cells in an animal model *PLoS ONE* **13** 1–13

[30] Lee Y K, Jung S K, Chang Y H and Kwak H S 2017 Highly bioavailable nanocalcium from oyster shell for preventing osteoporosis in rats *Int J Food Sci Nutr* **68** 931–40

[31] Ahmed S A, Gibriel A A Y, Abdellatif A K and Ebied H M 2015 Evaluation of food products fortified with oyster shell for the prevention and treatment of osteoporosis *J Food Sci Technol* **52** 6816–20

[32] Palaniappan M, Selvarajan S, Srinivasamurthy S and Chandrasekaran A 2014 Oyster shell calcium induced parotid swelling *J Pharmacol Pharmacother* **5** 256–7

[33] Lee H Z L, Davies I M, Baxter J M, Diele K and Sanderson W G 2020 Missing the full story: First estimates of carbon deposition rates for the European flat oyster, Ostrea edulis *Aquat Conserv Mar Freshw Ecosyst* **30** 2076–86

[34] Fodrie F J, Rodriguez A B, Gittman R K, Grabowski J H, Lindquist N L, Peterson C H, Piehler M F and Ridge J T 2017 Oyster reefs as carbon sources and sinks *Proc R Soc B Biol Sci* **284**

[35] Coen L D, Brumbaugh R D, Bushek D, Grizzle R, Luckenbach M W, Posey M H, Powers S P and Tolley S G 2007 Ecosystem services related to oyster restoration *Mar Ecol Prog Ser* **341** 303–7

[36] Gracia A, Rangel-Buitrago N, Oakley J A, Williams A T 2018 Use of ecosystems in coastal erosion management *Ocean Coast Manag* **156** 277–89

[37] Scyphers S B, Powers S P, Heck K L and Byron D 2011 Oyster reefs as natural breakwaters mitigate shoreline loss and facilitate fisheries *PLoS One* **6** e22396

[38] Cheong S M, Silliman B, Wong P P, Van Wesenbeeck B, Kim C K and Guannel G 2013 Coastal adaptation with ecological engineering *Nat Clim Chang* **3** 787–91

[39] Baggett L P, Powers S, Brumbaugh R, Coen L, DeAngelis B, Greene J, Hancock B and Morlock S 2014 *Oyster Habitat Restoration Monitoring and Assessment Handbook* (The Nature Conservancy)

[40] Grabowski J H and Peterson C H 2007 Restoring oyster reefs to recover ecosystem services *Ecosystem Engineers: Plants to Protists* (Academic Press) pp 281–98

[41] Newell R I E, Fisher T R, Holyoke R R and Cornwell J C 2005 Influence of Eastern Oysters on Nitrogen and Phosphorus Regeneration in Chesapeake Bay, USA *Comp Roles Suspens Ecosyst* **47** 93–120

[42] Cerco C F 2015 A Multi-module Approach to Calculation of Oyster (Crassostrea virginica) Environmental Benefits *Environ Manage* **56** 467–79

[43] Higgins C B, Stephenson K, Brown B L 2011 Nutrient Bioassimilation Capacity of Aquacultured Oysters: Quantification of an Ecosystem Service *J Environ Qual* **40** 271–7

[44] Troost K 2010 Causes and effects of a highly successful marine invasion: Case-study of the introduced Pacific oyster Crassostrea gigas in continental NW European estuaries *J Sea Res* **64** 145–65

[45] Campbell M D 2004 Analysis and evaluation of a bioengineered submerged breakwater. *LSU Master's Theses* 1220

[46] Walles B, Salvador de Paiva J, van Prooijen B C, Ysebaert T and Smaal A C 2015 The Ecosystem Engineer Crassostrea gigas Affects Tidal Flat Morphology Beyond the Boundary of Their Reef Structures *Estuaries and Coasts* **38** 941–50

[47] Lenihan H S, Micheli F, Shelton S W and Peterson C H 1999 The influence of multiple environmental stressors on susceptibility to parasites: An experimental determination with oysters *Limnol Oceanogr* **44** 910–24

[48] Farber S C, Costanza R and Wilson M A 2002 Economic and ecological concepts for valuing ecosystem services *Ecol Econ* **41** 375–92

[49] Fitzsimons E J *et al* 2019 *Restoration Guidelines for Shellfish Reefs* (Arlington VA, USA: The Nature Conservancy)

[50] Newell R I E and Koch E W 2004 Modeling Seagrass Density and Distribution in Response to Changes in Turbidity Stemming from Bivalve Filtration and Seagrass Sediment Stabilization *Estuaries and Coasts* **27** 793–806

[51] Donadi S, Van Der Heide T, Van Der Zee E M, Eklöf J S, Van De Koppe J, Weerman E J, Piersma T, Olff H and Eriksson B K 2013 Cross-habitat interactions among bivalve species control community structure on intertidal flats *Ecology* **94** 489–498

[52] Peterson B J and Heck K L 1999 The potential for suspension feeding bivalves to increase seagrass productivity *J Exp Mar Bio Ecol* **240** 37–52
[53] Wall C C, Peterson B J and Gobler C J 2008 Facilitation of seagrass Zostera marina productivity by suspension-feeding bivalves *Mar Ecol Prog Ser* **357** 165–74
[54] Pogoda B, Brown J, Hancock B, Preston J, Pouvreau S, Kamermans P, Sanderson W and Nordheim H V 2019 The Native Oyster Restoration Alliance (NORA) and the Berlin Oyster Recommendation: Bringing back a key ecosystem engineer by developing and supporting best practice in Europe *Aquat Living Resour* **32**
[55] Dasgupta P 2021 *The economics of biodiversity: The Dasgupta review* (London: HM Treasury)
[56] Pontee N, Narayan S, Beck M W and Hosking A H 2016 Nature-based solutions: Lessons from around the world *Marit Eng* **169** 29–36
[57] Potts T, Burdon D, Jackson E, Atkins J, Saunders J, Hastings E and Langmead O 2014 Do marine protected areas deliver flows of ecosystem services to support human welfare? *Mar Policy* **44** 139–48
[58] Stine E and Luna-Mega C 2019 Musical Aesthetics of the Natural World: Two Modern Compositional Approaches *Jefferson J Sci Cult* **5** 21–32
[59] Billion Oyster Project 2021 STEM Education (Online) Available from: https://www.billionoysterproject.org/
[60] zu Ermgassen P S E *et al* 2012 Historical ecology with real numbers: Past and present extent and biomass of an imperilled estuarine habitat *Proc R Soc B Biol Sci* **279** 3393–400
[61] Kroeger T 2012 Dollars and Sense: Economic Benefits and Impacts from two Oyster Reef Restoration Projects in the Northern Gulf of Mexico (The Nature Conservancy)
[62] Gercken J and Schmidt A 2014 *Current Status of the European Oyster (Ostrea edulis) and Possibilities for Restoration in the German North Sea* (Neu Broderstorf, Germany: Bundesamt für Naturschutz)
[63] Lallias D, Boudry P, Lapègue S, King J W and Beaumont A R 2010 Strategies for the retention of high genetic variability in European flat oyster (Ostrea edulis) restoration programmes *Conserv Genet* **11** 1899–910
[64] Culloty S C and Mulcahy M F 2007 Bonamia Ostreae in the Native Oyster Ostrea Edulis *Mar Environ Heal Ser* **29** 1–36
[65] Hudson E B and Hill B J 1991 Impact and spread of bonamiasis in the UK *Aquaculture* **93** 279–85
[66] Carral L, Fabal C C, Galdo M I L, Rodríguez-Guerreiro M J and Barros J J C 2020 Assessment of the materials employed in green artificial reefs for the galician estuaries in terms of circular economy *Int J Environ Res Public Health* **17** 1–24
[67] MARINEFF 2020 MARINEFF (MARine INFrastructure EFFects) Newsletter 4 (Online) Available from: http://marineff-project.eu/wp-content/uploads/2020/05/May-2020-Newsletter_CIRIA-1.pdf
[68] Harding S, Nelson L, Glover T 2016 Solent Oyster Restoration Project Management Plan (Blue Marine Foundation)
[69] Sas H *et al* 2020 Bonamia infection in native oysters (Ostrea edulis) in relation to European restoration projects *Aquat Conserv Mar Freshw Ecosyst* **30** 2150–62

Primary production assessment on eco-engineering infrastructures: English Channel case study

P Claquin[1,2,3], **B Vivier**[1,2,3], **M Navon**[1,2,3], **A M Rusig**[1,2,3], **I Mussio**[1,2,3], **L Chasselin**[1,3], **M Deloor**[1,2,3], **J C Dauvin**[1,4], **M Boutouil**[5], **F Orvain**[1,2]

[1] Normandie Université, Université de Caen Normandie, F-14032, Caen, France
[2] Laboratoire Biologie des ORganismes et Ecosystèmes Aquatiques (BOREA, UMR CNRS 8067), Muséum National d'Histoire Naturelle, Sorbonne Université, Université de Caen Normandie, IRD 207, Université des Antilles. Centre de Recherches en Environnement Côtier (CREC) - Station Marine, BP49, 54, rue du Docteur Charcot - 14530 Luc-sur-Mer, France
[3] Centre de Recherches en Environnement Côtier (CREC) - Station Marine de l'Université de Caen Normandie, BP49, 54, rue du Docteur Charcot - 14530
[4] Laboratoire Morphodynamique Continentale et Côtière, UMR CNRS 6143 M2C, Normandie Université, Université de Caen Normandie, UNIROUEN, 24 rue des Tilleuls, F-14000 Caen, France
[5] Ecole Supérieure d'Ingénieurs des Travaux de la Construction de Caen (ESITC Caen), 14610 Epron, France

Email: pascal.claquin@unicaen.fr

Abstract. In the Marineff project, experimental eco-engineering dyke modules were immersed in 2020 in the Bay of Cherbourg (English Channel). Eco-engineering and classic dyke modules were compared regarding biodiversity and productivity. This paper presents the monitoring strategy of primary production assessment applied. Primary production assessment was performed by coupling various technics in order to consider all the compartments of primary producers present on the dykes and in the associated surrounding water mass. Seaweeds, microphytobenthos and phytoplankton were studied. Benthic chambers measuring oxygen and pCO_2 under controlled light, and technics based on variable fluorescence were applied on eco-engineering and classic dyke modules. Preliminary results are presented in order to illustrate the expected indicator and methodology improvement and limits were discussed. The whole data set will allow us to explore the relationship between productivity and biodiversity as a function of the sites and of the nature of the infrastructure, eco-engineering or not. At this stage of our study we pointed out that the macroalgae were 7.9 times more productive than biofilm per surface on eco-engineering module whilst only 1.5 times on classic module. Smooth surface appeared to favour biofilm productivity at this stage of colonisation while rough surface drove more quickly the system to a production dominated by macroalgae.

1. Introduction

Coastal artificialisation is a widespread phenomenon that leads to irreversible loss in many ecosystems. Artificialisation is a highly visible footprint of anthropic activities related to commercial exchanges, energy production, fisheries, tourism, leisure activities, etc. A reduction of this footprint is required but is socially and economically complex [1, 2]. Marine eco-engineering infrastructures can limit negative effects and enhance the ecological status of artificialized areas [2]. Artificialisation of shallow coastal ecosystems by marine infrastructures offers new substrates for benthic microalgae and macroalgae which may increase local primary production [1]. Primary production is at the base of trophic network and drives many chemical cycles at local and global scales [3-6]. Oxygenic primary production, which represents the production of chemical energy in organic compounds by photosynthetic organisms is one of the major ecological functions of ecosystem in photic area [7, 8]. Macroalgae (i.e. seaweeds) which belong to three phyla or classes, Ulvophyceae (Chlorophyta - green algae), Rhodophyta (red algae) and Phaeophyceae (Ochrophyta - brown algae); microalgae which regroups 10 eukaryotic phyla including various taxa like diatoms, dinoflagellates, Prymnesiophyceae [9], seagrasses (Angiosperm) and Cyanobacteria which belong to prokaryotes represent a very large biodiversity of primary producers. The functioning of these ecosystems and the interaction between benthic flora and fauna influence also the phytoplankton of water masses surrounding [2, 10]. An assessment of the whole primary production related to artificial infrastructure is then required to characterize the functioning and the ecological trajectories of these new ecosystems [1, 2]. Methods allowing primary production measurements of the different components of primary producers have to be deployed in parallel. Several technics based on O_2, CO_2, carbon isotopes, or fluorescence variations measurements were commonly used to assess primary production [11, 12]. Coupling different technics allow to consider various organisms and processes as a function of the primary producer size and time scale. In this context, marine eco-engineering infrastructures were developed to improve ecological impacts of artificialized areas [1, 2]. In the European INTERREG Va MARINEFF project, experimental eco-engineering dyke modules and classic modules were immersed in September 2020 in the bay of Cherbourg and a monitoring of primary production dynamics was initiated. In the present paper the selected main protocols are briefly described and preliminary results are presented and discussed.

2. Materials & methods

2.1. Sampling site

Twelve dyke modules were immerged into 10 m depth on sandy soft bottom in the bay of Cherbourg in September 2020. Two types were deployed: six classic modules corresponding to concrete cubes of 1.7 m^3 with smooth external texture (Figure 1a) and six eco-engineering modules corresponding to concrete cube of 1.7 m^3 with rough external texture and various holes, gutters, overhangs, grooves of various sizes and positions to promote biological colonisation (Figure 1b, Figure 2). Eco-engineering modules were designed by MARINEFF project patterns (University of Caen Normandie, TPC - VINCI Construction France, and ESITC) and produced by TPC - VINCI Construction France.

a b

Figure 1. Dyke modules deployed during the MARINEFF project: a: classic module; b: Eco-engineering module.

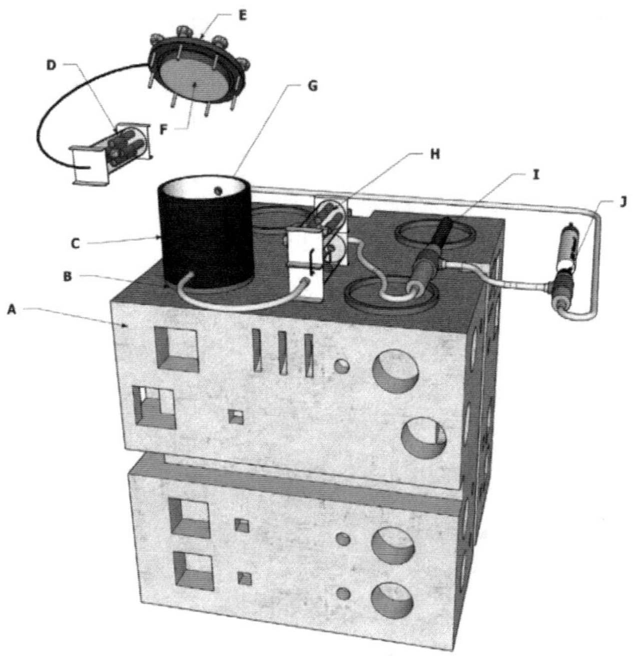

Figure 2. Automated benthic chamber for primary production measurement. (A) MARINEFF eco-engineering dyke module with the mooring ring (B) to fix the benthic chamber (C). Aluminium tape (G) covers the benthic chamber inside. LED panel (F) is fixed to the benthic chamber top (E) which is screwed on the benthic chamber to close it hermetically. The LED panel is driven by the LED controller (D). Water flow is generated by an underwater pump (H). Two sensors, YSI EXO 1 (I) with oxygen probe, and Turner Design C-Sense pCO2 probe (J) driven by the RBR Virtuoso3 logger (J), measures the gas concentration. The whole system is in a water closed circuit connected by pipe.

2.2. Primary production assessments (Figure 3)

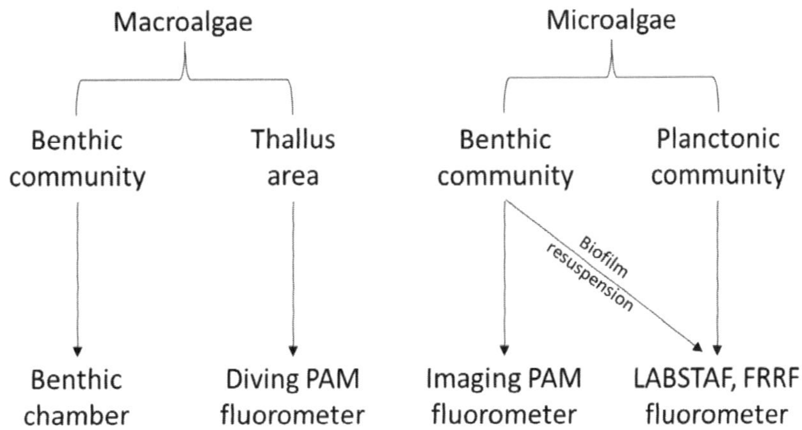

Figure 3. Monitoring strategy deployed for primary production assessment on eco-engineering infrastructures during the MARINEFF project.

2.2.1. *Benthic chamber.* A fully automated system for measurements of photosynthetic oxygen and CO_2 exchange was developed in the MARINEFF project (Figure 2). The benthic chamber allows generating light gradient between 0 to 900 µmol photons m^{-2} s^{-1} produced by white LEDs with maximum pics at 450 nm and 600 nm (Figure 4). Light intensity and the duration of each light step are controlled by a programmable controller unit. Photoactive radiation (PAR) intensity of each step was measured using a RBR solo3 PAR logger connected to a Li-COR "Underwater Quantum Sensor" LI-192. Dissolved oxygen was recorded with an EXO Optical Dissolved Oxygen Sensor and pCO2 by a Turner Designs C-sense pCO_2 probe every 30 seconds. Temperature and pH were measured at the same frequency in the chamber. A pump guarantied the water homogenisation during the incubations. In the present experiment, 7 light steps of 15 min each from 0 to 750 µmol photons m^{-2} s^{-1} were applied. A dark phase was applied after the last light step in order to estimate the respiration rate. Because O_2 evolution consumption in the light, may differ from dark O_2 consumption, the respiration rate was estimated during 5 min just after the last light phase [13]. O_2 measurements were used to estimate respectively net primary production (NPP) and benthic community respiration (BCR) of the phyto- and zoo-benthos community [14]. GPP was then calculated from NPP and BCR measurements (GPP = NPP + BCR). pCO_2 measurements have to be associated to alkalinity and/or DIC measures at the beginning and at the end of incubations to be properly exploited [15]. This step of the experiments is not presented in this paper.

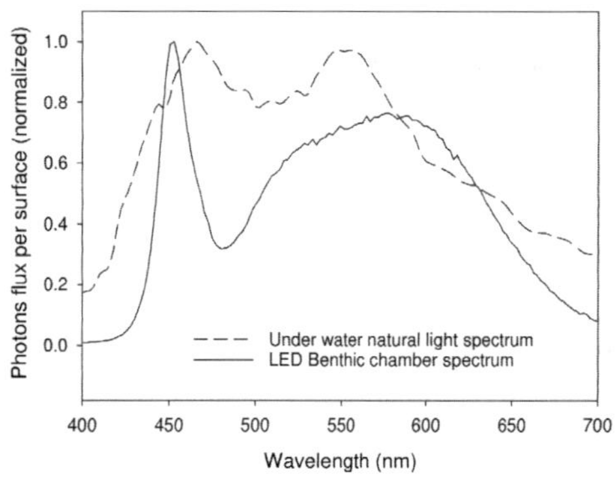

Figure 4. LED spectrum used in the automated benthic chamber and under water natural light spectrum.

2.2.2. Variable fluorescence measurements

2.2.2.1. *Chlorophyll- a fluorescence induction pulse amplitude modulation (PAM).* Diving and imaging PAM (WaltzTm, Effeltrich, Germany) are used in the assessment protocol. Imagining PAM is not present in this paper. See Vivier et al. [16] for more details. The maximum quantum yield (Fv/Fm in the dark) and the effective photochemical efficiency (Fq'/Fm' in the light), were measured on the module by scuba divers on several species of macroalgae using a Diving PAMII (WaltzTm, Effeltrich, Germany) [17].The macroalgae sample placed in the dark was excited by a weak blue measuring light (1 µmol m^{-2} s^{-1}, 474 nm, frequency 0.6 kHz) to record minimum fluorescence (F_0). The optical fiber of the PAM was positioned exactly 5 mm from the surface of the macroalgae. Maximum fluorescence (Fm) was measured using a multiple turnover flash method [17]. We used a saturating blue light pulse (0.6 s, 4000 µmol m^{-2} s^{-1} at 474 nm) produced by a LED allowing reduction of the quinone A (QA), quinone B (QB) and part of plastoquinone (PQ) pools.

Fv/Fm was calculated according to the following equation [18]:

$$\frac{F_v}{F_m} = \frac{(F_m - F_0)}{F_m}$$

The samples were exposed to 10 irradiances (E) from 0 to 1000 µmol photons m^{-2} s^{-1} at 474 nm for 20 seconds at each step. Steady state fluorescence (Fs) and maximum fluorescence (Fm') were measured. The effective quantum efficiency of PSII for each irradiance was determined as follows [18]:

$$\frac{\Delta F}{F_{m'}} = \frac{(F_{m'} - F_s)}{F_{m'}}$$

The electron transport rate (ETR, µmol electrons m^{-2} s^{-1}) was calculated for each irradiance as described by Figueroa et al. [19]. ETR is a measure of the rate of linear electron transport through photosystem II, which is correlated with the overall photosynthetic performance of the algae. In order to properly estimate the ETR, light absorption (A) has to be measured in laboratory on sample collected on dyke modules.

$$ETR(E) = \mathrm{AQ}_\lambda \times FII \times \frac{\Delta F}{F_{m'}}$$

Where AQ_λ is the absorbed quanta calculated as the product of the integration of the spectral absorptance (A_λ) between 400-700 nm and spectral irradiance of the light source (E_λ), FII is the fraction of AQ directed to PS II including its light harvesting complexes. FII for Rhodophyta is about 0.15 and 0.5 for Ulvophyceae and Phaeophyceae [19-21]. The absorptance (A_λ) was determined for each 0.5 nm between 400 and 700 nm by means of an integrating sphere installed in a Shimadzu UV2600i spectrophotometer [19]. Fluorescence excitation spectra from 400 to 700 nm with emission measured at 730 nm was performed using a Shimadzu Spectrofluorometer RF-6000 [19].

2.2.2.2. *LabSTAF fluorometer.* Photosynthetic parameters of the phytoplankton and of the resuspended microalgal biofilm were determined by using a LabSTAF (Chelsea Technologies, UK) fluorimeter. The microalgal biofilm was collected by scuba divers using a syringe of 1.66 L specially designed for this sampling. 25 cm^2 were resuspended using toothbrush and drawn into the syringe. The LabSTAF allows measurements at different wavelengths. A single turnover (ST) saturation phase is delivered with a solid flash if 100 µs to measure minimal and maximum fluorescence as described in Boatman et al. (2019) [22]. Samples were exposed to 12 intensities of increasing PAR (0 – 1,500 µmol photon. $s^{-1}.m^{-2}$) with 30-second intervals in order to obtain fluorescent-light curve (FLC). The Labstaf enables to measure the electron flux from the PSII per volume unit (JV_{PII}, mmol electrons m^{-3} h^{-1}) which allowed to estimate the oxygen production (GO_{PII}) in mmol O_2 m^{-2} h^{-1}.

$$JV_{PII} = a_{LHII} \times rETR$$

With a_{LHII} the absorption coefficient of PSII in m^{-1} (see Boatman et al. (2019) [22] for details). The LabSTAF also allows to measure the absorption cross section of PSII, $Sigma_{PSII}$ ($nm2.PSII^{-1}$) at different wavelengths. The same measurements were performed on water samples around modules in order to estimate the phytoplankton production. A correction was then applied on microalgal biofilm measurements which are slightly contaminated by phytoplankton during sampling.

2.2.3. Production / Light curves. The GPP, ETR and GO_{PII} were plotted against light (PAR) produced respectively by the benthic chamber, the Diving PAM and the LabSTAF. To estimate photosynthetic parameters, the mathematical model of Webb et al. (1974) [23] was applied to the data:

$$X(PAR) = Xmax \times e^{(1-\frac{PAR \times \alpha_X}{Xmax})}$$

where X (PAR) is GPP (PAR), ETR (PAR) or GO_{PII} (PAR). Xmax is the maximum photosynthetic capacity measured with the O_2 incubation method (GPP_{max} in mmol O_2 m^{-2} h^{-1}), the PAM method (ETRmax in µmol electrons m^{-2} s^{-1}) or with LabSTAF (GO_{PIImax}, mmol O_2 m^{-2} h^{-1}), α is the photosynthetic efficiency which corresponds to the slope of each Production / Light curve.

3. Results & Discussion

This paper presents the strategy deployed to measure the primary production of the different component of artificial infrastructure installed in a photic zone of a macrotidal epicontinental sea. The benthic chamber which was developed in this context allowed to standardize light exposure of the macroalgae community during incubation. Traditionally transparent benthic chambers were used [24] and the results obtained were highly dependent on the daily irradiance which could change from one day to the next and between seasons [14, 24]. The automated benthic chamber used in this study is easily installed by scuba divers and enabled reproductible experiments regarding light exposure. The figures 5a and 5b showed that light energy of the benthic chamber allowed to reach a saturation of GPP which was required to properly fit a Webb model in order to estimate the GPPmax and the photosynthetic efficiency alfa. This preliminary result showed a higher maximal production rate (GPP_{max} = 14.18 ± 2.19 mmol O_2 m^{-2} h^{-1}) and a higher alfa (0.12 ± 0.03 mmol O_2 m^{-2} h^{-1} (µmol photons m^{-2} s^{-1})$^{-1}$) for Eco-engineering modules than for classic module (GPP max = 8.88 ± 0.62 mmol O_2 m^{-2} h^{-1}; alfa = 0.06 ± 0.008 mmol O_2 m^{-2} h^{-1} (µmol photons m^{-2} s^{-1})$^{-1}$) (Figure 5a). Biomass measurements confirmed a higher covering on eco-engineering modules (405.8 g m^{-2}) than on classic modules (256.0 g m^{-2}). The productivity per biomass unit is comparable between both types of modules, with GPP_{max} per weight of 34.96 ± 6.04 µmol O_2 g^{-1} h^{-1} and alfa 0.32 ± 0.09 for eco-engineering modules and GPP_{max} per weight of 34.63 ± 2.44 µmol O_2 g^{-1} h^{-1} and alfa 0.25 ± 0.03 for classic modules (Figure 5b). The equal value of productivity firstly validated our protocol with this new benthic chamber and secondly showed that the advantage of engineering modules is related to fixation success of macroalgae and on the efficiency of the colonisation on rough substrate instead of smooth one [25]. The measurements using benthic chambers will be performed 4 times per year in triplicate on both module types.

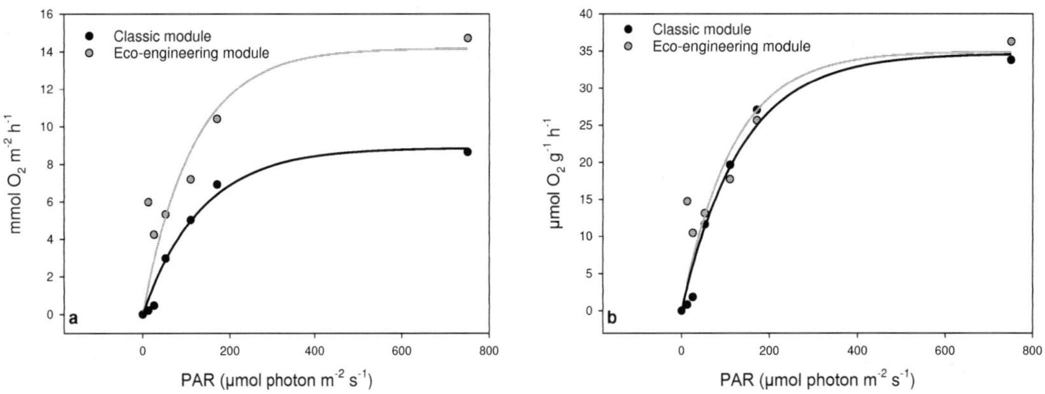

Figure 5. Example of Production / Light curves measured on MARINEFF eco-engineering dyke module and classic dyke module in the Bay of Cherbourg in April 2021. Dots show the GPP measurements at each light intensity (PAR) and the lines show the Webb model fits. a: GPP expressed as a function of subtract surface and time; b: GPP expressed as a function of fresh weight of macroalgae and time.

In addition to benthic chamber experiments, diving PAM measurements were used to estimate photosynthetic parameters *in situ*. GPP estimation gave a production capacity of the whole community while PAM measurements allowed to measure photosynthetic parameters and productivity capacity individually for each species [26] and/or for different parts of thallus for massive species. PAM techniques provide a non-destructive way of assessing the photo-physiology of a range of organisms, including phytoplankton, macroalgae and higher plants. A number of works have used chlorophyll- a fluorescence techniques to study the photosynthetic characteristics of macroalgae, both in the field and in the laboratory [21, 27-30]. In the present study, several species of green, brown and red algae were measured (*Saccorhiza polyschides, Dictyota dichotoma, Gracilaria gracilis, Palmaria palmata, Cryptopleura ramosa, Ulva lactuca, Bryopsis plumosa*) and regrouped as a function of their phylum. Comparison of PAM measurements from different species should be conducted with care, as the irradiance absorption of a photosynthetic organism may change, affecting ETR. As shown in the example presented in figure 6, absorption spectra are highly variable as a function of pigment composition. Absorption spectrum are then required to properly estimate ETR as presented in figure

7a. This absorption spectrum can be improved by using excitation spectrum measurements [19]. According to the literature FII can vary between 0.15 for red algae to 0.5 for green algae [30]. ETR and alfa were significantly higher in brown algae than in red and green. The data were spectrally corrected and brown algae showed a high absorption at 474 nm with high efficiency for the major light-harvesting complex of photosystem II at this wavelength while red algae had a poor light transfer to PSII at this wavelength.

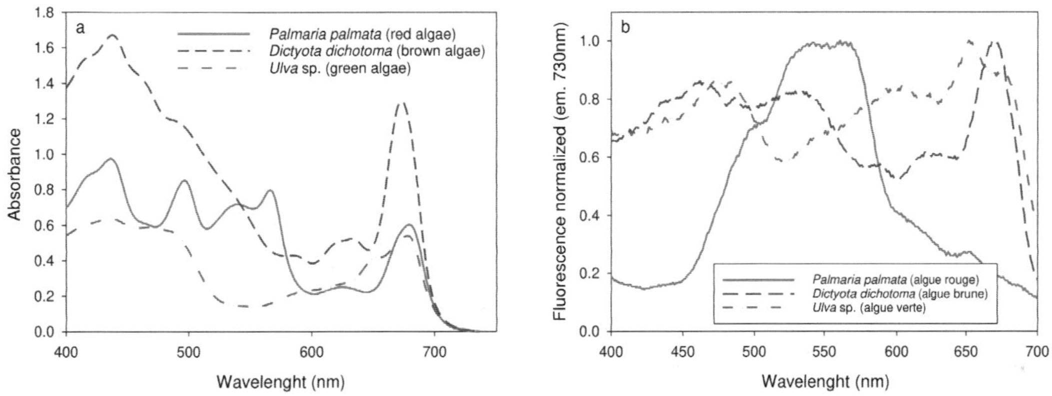

Figure 6. a: Example of spectral absorptance (A_λ) between 400-700 nm and b: Example of fluorescence excitation spectra between 400-700 nm with an emission measured at 730 nm performed on Chlorophyta (green algae), Rhodophyta (red algae) and Phaeophyceae (brown algae) from the Bay of Cherbourg.

The value of ETR are in accordance with literature [30] but calculation of this parameter is very heterogeneous in literature. Beside relative ETR (rETR) [31] measurements, calculation of ETR per unit surface as a function of time are conducted in different ways in literature [26, 27, 30, 32]. The way how to calculate the absorption can change or a standard factor can be applied [32], FII can be considered as variable as a function of pigment groups ([17, 19, 20, 30] or constant for all groups [26, 32]. According to our knowledge the approach proposed by Figueroa et al [19] followed the same principle as described by Johnsen & Sakshaug [33] for phytoplankton. This method which required a determination of absorption factor using an integrative sphere and a determination of FII by using spectrofluorimetry measurements will be applied in our monitoring strategy. Beside producing indicator, such an approach allows to explore ecophysiological regulation of light absorption as a function of environmental parameters. The Fv/Fm obtained for green and brown algae are found in healthy algae, well acclimated to their environment. The lowest value obtained on red algae is probably partly due to pigment composition and is in accordance with literature [32]. The response of red algae to PAM fluorometry techniques can be different from responses in other algae groups due to the presence of phycobilisomes in the photosynthetic apparatus [34, 35]. State transition mechanisms frequently observed in red algae can influence the Fv/Fm [32, 36-38]. We did not exempt to have differences on photosynthetic parameters between the same species collected from eco-engineering modules or classic modules at this stage of the monitoring. If biomass remained significantly higher for longer, the shading due to algae canopy would lead to a discrepancy of algae photoacclimation and possibly of biodiversity and community structures between both types of module.

The microalgal biofilm was also studied in this project. Colonisation of hard substrates by microalgal biofilm is a crucial step for the ecological succession of new hard substrata [16, 39]. Imaging PAM is used to follow the colonisation on small samples of module fixed on dyke modules which are regularly collected [16]. This approach is not detailed in the present paper. As described in the Material and Method section, 25 cm^2 of biofilm were resuspended and analysed using a LabSTAF (Figure 7b).

Contrary to macroalgae, the biofilm of classic modules was more productive. The GO_{IImax}, which correspond to GPP max of macroalgae, was 5.89 ± 0.65 mmol O_2 m^{-2} h^{-1} for classic module and 1.79 ± 1.07 mmol O_2 m^{-2} h^{-1} for eco-engineering module. Biofilm from classic modules also presented a higher Fv/Fm than biofilm from eco-engineering modules. We pointed out that de absorption cross-section of the PSII was also different, which indicated variation of photoacclimatation process or difference on community structure [40, 41]. Analysis of the biofilm community structure is still in process by coupling microscopic and molecular approaches. The higher biomass observed on eco-engineering module probably influenced the biofilm production capacities and its acclimation as a function of light and nutrients. The interactions between microalgal biofilm and macroalgae as well as the temporal variations on these interactions will be a major point of the future data analyses. These first data allowed to compare the primary production of macroalgae and microalgal biofilm. The macroalgae were 7.9 times more productive than biofilm per surface on eco-engineering module whilst only 1.5 times on classic module. Smooth surface appeared to favour biofilm productivity at this stage of colonisation while rough surface drove more quickly the system to a production dominated by macroalgae.

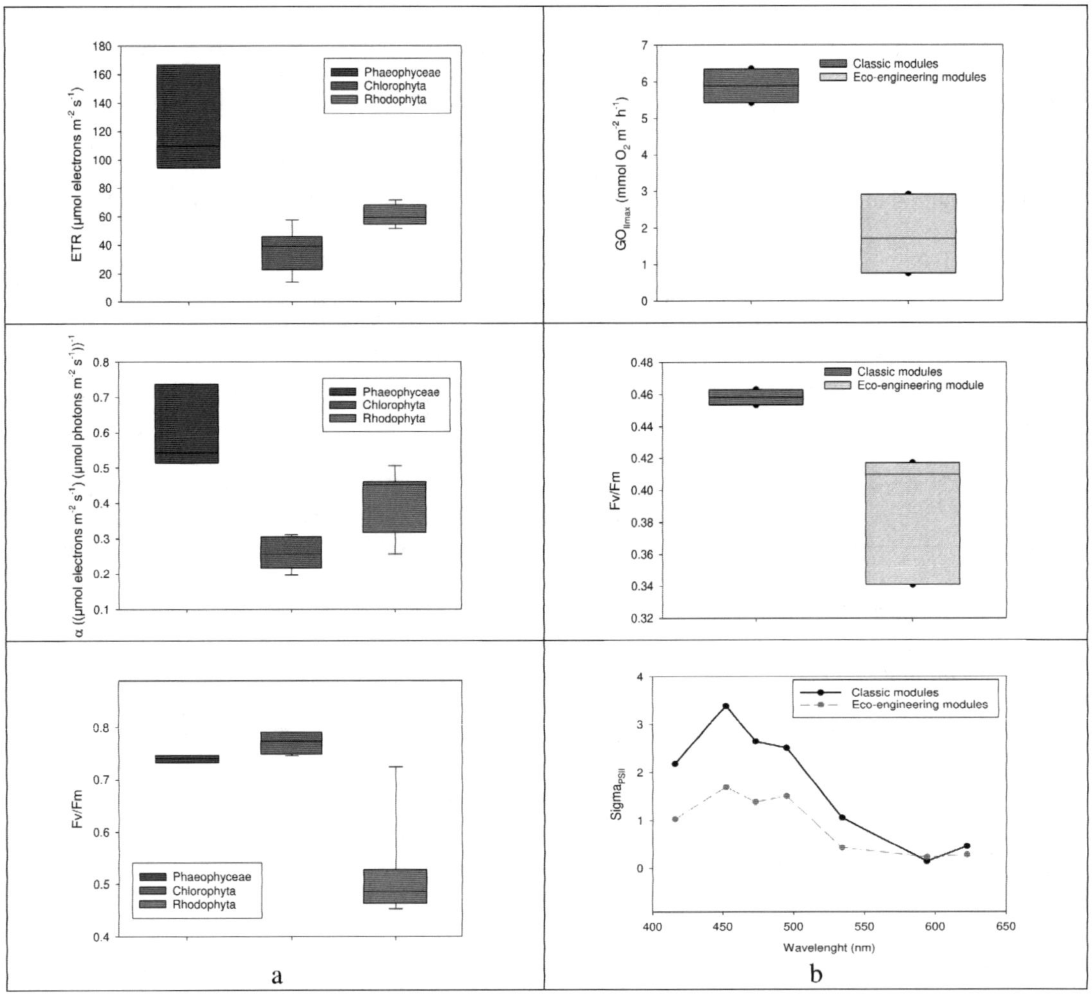

Figure 7. a: ETR, alfa and Fv/Fm measured under water by using Diving PAMII on the dykes modules on Chlorophyta (green algae), Rhodophyta (red algae) and Phaeophyceae (brown algae). b: GO_{IImax},

Fv/Fm, and Sigma PSII measured on microalgal biofilm by using LabSTAF. Measurements were performed in the Bay of Cherbourg in April 2021

4. Conclusion

At the preliminary stage of our monitoring, the sampling and lab protocols deployed were validated. The selection of indicators and their metrics have to be confirmed and some complementary data are required to properly interpret all these first data. We observed differences regarding macro and microalgae primary production. There was no significant difference between productivity of macroalgae (GPP.g^{-1}.h^{-1}) measured on eco-engineering and classic modules (ANOVA, $p = 0.967$), because productivity is controlled by environmental parameters like nutrients and light availability while biomass beside productivity capacities also depends on the rate of fixation efficiency. The higher biomass observed on eco-engineering led to a higher GPP per surface unit. For equal surface, at this preliminary stage of our study, eco-engineering modules appeared more productive that classic modules. The monitoring will allow us to increase the replication of these experiments and to confirm or not this trajectory for longer period. The colonization surface of classic module may catch up after several months or years. Competition for space with macrofauna will be also an important point to consider in our monitoring [42]. The primary production estimations will be used in an ecological model. The approach exposed in this work will also allow providing guidelines and efficient protocols to assess primary production in artificial structures and to produce indicators of this main ecology function.

Acknowledgments

The results were obtained in the collaborative project Interreg Va MARINEFF, selected under the European Cross-border Program INTERREG FMA, and co-funded by the ERDF. The authors wish to thank the co-financiers and all project partners for their support. The authors thank the Normandy Region for funding the IMAGING-PAM and LabSTAFF (CPER Manche 2021). The authors thank D Lemeille and J P Lehodey from the CREC-UniCaen for their technical supports.

References

[1] Vivier B, Dauvin J-C, Navon M, Rusig A-M, Mussio I, Orvain F, Boutouil M and Claquin P 2021 Marine artificial reefs, a meta-analysis of their design, objectives and effectiveness *Global Ecology and Conservation* **27** e01538

[2] Layman C A and Allgeier J E 2020 An ecosystem ecology perspective on artificial reef production *J. Appl. Ecol.* **57** 2139-48

[3] Cloern J E, Foster S Q and Kleckner A E 2014 Phytoplankton primary production in the world's estuarine-coastal ecosystems *Biogeosciences* **11** 2477-501

[4] Underwood G J C and Kromkamp J 1999 *Advances in Ecological Research, Vol 29: Estuaries,* ed D B Nedwell and D G Raffaelli pp 93-153

[5] Falkowski P G, Barber R T and Smetacek V 1998 Biogeochemical controls and feedbacks on ocean primary production *Science* **281** 200-6

[6] Field C B, Behrenfeld M J, Randerson J T and Falkowski P 1998 Primary production of the biosphere: Integrating terrestrial and oceanic components *Science* **281** 237-40

[7] Cardinale B J, Matulich K L, Hooper D U, Byrnes J E, Duffy E, Gamfeldt L, Balvanera P, O'Connor M I and Gonzalez A 2011 The functional role of producer diversity in ecosystems *American Journal of Botany* **98** 572-92

[8] Kemp W M, Boynton W R, Adolf J E, Boesch D F, Boicourt W C, Brush G, Cornwell J C, Fisher T R, Glibert P M, Hagy J D, Harding L W, Houde E D, Kimmel D G, Miller W D, Newell R I E, Roman M R, Smith E M and Stevenson J C 2005 Eutrophication of Chesapeake Bay: historical trends and ecological interactions *Mar. Ecol. Prog. Ser.* **303** 1-29

[9] Adl S M, Simpson A G B, Lane C E, Lukes J, Bass D, Bowser S S, Brown M W, Burki F, Dunthorn M, Hampl V, Heiss A, Hoppenrath M, Lara E, le Gall L, Lynn D H, McManus H, Mitchell E A D, Mozley-Stanridge S E, Parfrey L W, Pawlowski J, Rueckert S, Shadwick L,

Schoch C L, Smirnov A and Spiegel F W 2012 The Revised Classification of Eukaryotes *Journal of Eukaryotic Microbiology* **59** 429-93

[10] Zhu W, Qin C, Ma H, Xi S, Zuo T, Pan W and Li C 2020 Response of protist community dynamics and co-occurrence patterns to the construction of artificial reefs: A case study in Daya Bay, China *Science of the Total Environment* **742**

[11] Falkowski P G and Raven J A 1997 *Aquatic Photosynthesis* (Malden, USA: Blackwell Science)

[12] Napoleon C and Claquin P 2012 Multi-Parametric Relationships between PAM Measurements and Carbon Incorporation, an In Situ Approach *PLoS One* **7**

[13] Claquin P, Kromkamp J C and Martin-Jézéquel V 2004 Relationship between photosynthetic metabolism and cell cycle in a synchronized culture of the marine alga *Cylindrotheca fusiformis* (Bacillariophyceae) *Eur. J. Phycol.* **39** 33-41

[14] Hubas C and Davoult D 2006 Does seasonal proliferation of Enteromorpha sp affect the annual benthic metabolism of a small macrotidal estuary? (Roscoff Aber Bay, France) *Estuarine Coastal and Shelf Science* **70** 287-96

[15] Abril G 2009 Comments on: "Underwater measurements of carbon dioxide evolution in marine plant communities: A new method" by J. Silva and R. Santos Estuarine, Coastal and Shelf Science 78(2008) 827-830 Discussion *Estuarine Coastal and Shelf Science* **82** 357-60

[16] Vivier B, Claquin P, Lelong C, Lesage Q, Peccate M, Hamel B, Georges M, Bourguiba A, Bourguiba N, Boutouil M, Goux D, Dauvin J C and Orvain F Submitted Influence of biomaterial composition and microtopography on marine biofilms growth and photobiology *Biofouling*

[17] Schreiber U, Gademann R, Ralph P J and Larkum A W D 1997 Assessment of photosynthetic performance of Prochloron in Lissoclinum patella in hospite by chlorophyll fluorescence measurements *Plant and Cell Physiology* **38** 945-51

[18] Genty B, Briantais J M and Baker N R 1989 The relationship between the quantum yield of photosynthetic electron-transport and quenching of chlorophyll fluorescence *Biochim. Biophys. Acta* **990** 87-92

[19] Figueroa F L, Escassi L, Perez-Rodriguez E, Korbee N, Giles A D and Johnsen G 2003 Effects of short-term irradiation on photoinhibition and accumulation of mycosporine-like amino acids in sun and shade species of the red algal genus Porphyra *Journal of Photochemistry and Photobiology B-Biology* **69** 21-30

[20] Figueroa F L, Conde-Alvarez R and Gomez I 2003 Relations between electron transport rates determined by pulse amplitude modulated chlorophyll fluorescence and oxygen evolution in macroalgae under different light conditions *Photosynth. Res.* **75** 259-75

[21] Schneider G, Figueroa F L, Vega J, Aviles A, Chaves P, Horta P A, Korbee N and Bonomi-Barufi J 2020 Physiological and biochemical responses driven by different UV-visible radiation in Osmundea pinnatifida (Hudson) Stackhouse (Rhodophyta) *Photochem. Photobiol. Sci.* **19** 1650-64

[22] Boatman T G, Geider R J and Oxborough K 2019 Improving the Accuracy of Single Turnover Active Fluorometry (STAF) for the Estimation of Phytoplankton Primary Productivity (PhytoPP) *Frontiers in Marine Science* **6**

[23] Webb W L, Newton M and Starr D 1974 Carbon-dioxide exchange of alnus-rubra - Mathematical-model *Oecologia* **17** 281-91

[24] Gevaert F, Delebecq G, Menu D and Brutier L 2011 A fully automated system for measurements of photosynthetic oxygen exchange under immersed conditions: an example of its use in Laminaria digitata (Heterokontophyta: Phaeophyceae) *Limnology and Oceanography-Methods* **9** 361-79

[25] Schneck F, Schwarzbold A and Melo A S 2011 Substrate roughness affects stream benthic algal diversity, assemblage composition, and nestedness *J. N. Am. Benthol. Soc.* **30** 1049-56

[26] Migne A, Delebecq G, Davoult D, Spilmont N, Menu D and Gevaert F 2015 Photosynthetic activity and productivity of intertidal macroalgae: In situ measurements, from thallus to community scale *Aquatic Botany* **123** 6-12
[27] Beer S and Axelsson L 2004 Limitations in the use of PAM fluorometry for measuring photosynthetic rates of macroalgae at high irradiances *Eur. J. Phycol.* **39** 1-7
[28] Figueroa F L, Salles S, Aguilera J, Jimenez C, Mercado J, Vinegla B, FloresMoya A and Altamirano M 1997 Effects of solar radiation on photoinhibition and pigmentation in the red alga Porphyra leucosticta *Mar. Ecol. Prog. Ser.* **151** 81-90
[29] Franklin L A and Badger M R 2001 A comparison of photosynthetic electron transport rates in macroalgae measured by pulse amplitude modulated chlorophyll fluorometry and mass spectrometry *J. phycol.* **37** 756-67
[30] Figueroa F L, Hermoso-Beltran M, Celis-Pla P S M, Bonomi-Barufi J, Alvarez-Gomez F, Korbee N, Lopez-Calderon J, Cota-Lucero T and Riosmena-Rodriguez R 2016 Photosynthetic activity estimated as in vivo chlorophyll a fluorescence in calcareous red macroalgae *Ciencias Marinas* **42** 139-55
[31] Ralph P J and Gademann R 2005 Rapid light curves: A powerful tool to assess photosynthetic activity *Aquatic Botany* **82** 222-37
[32] Kim J H, Choi S J and Lee S 2019 Effects of temperature and light on photosynthesis and growth of red alga Pyropia dentata (Bangiales, Rhodophyta) in a conchocelis phase *Aquaculture* **505** 167-72
[33] Johnsen G and Sakshaug E 2007 Biooptical characteristics of PSII and PSI in 33 species (13 pigment groups) of marine phytoplankton, and the relevance for pulse-amplitude-modulated and fast-repetition-rate fluorometry *J. phycol.* **43** 1236-51
[34] Kana R, Kotabova E, Lukes M, Papacek S, Matonoha C, Liu L-N, Prasil O and Mullineaux C W 2014 Phycobilisome Mobility and Its Role in the Regulation of Light Harvesting in Red Algae *Plant Physiol.* **165** 1618-31
[35] Kowalczyk N, Rappaport F, Boyen C, Wollman F-A, Collen J and Joliot P 2013 Photosynthesis in Chondrus crispus: The contribution of energy spill-over in the regulation of excitonic flux *Biochimica Et Biophysica Acta-Bioenergetics* **1827** 834-42
[36] Allen J F 2003 State transitions - a question of balance *Science* **299** 1530-2
[37] Williams W P and Allen J F 1987 State-1/state-2 changes in higher-plants and algae *Photosynth. Res.* **13** 19-45
[38] Gefen-Treves S, Kedem I, Weiss G, Wagner D, Tchernov D and Kaplan A 2020 Acclimation of a rocky shore algal reef builder Neogoniolithon sp. to changing illuminations *Limnol. Oceanogr.* **65** 27-36
[39] Ly O, Yoris-Nobile A I, Sebaibi N, Blanco-Fernandez E, Boutouil M, Castro-Fresno D, Hall A E, Herbert R J H, Deboucha W, Reis B, Franco J N, Borges M T, Sousa-Pinto I, van der Linden P and Stafford R 2021 Optimisation of 3D printed concrete for artificial reefs: Biofouling and mechanical analysis *Construction and Building Materials* **272**
[40] Serre-Fredj L, Jacqueline F, Navon M, Izabel G, Chasselin L, Jolly O, Repecaud M and Claquin P 2021 Coupling high frequency monitoring and bioassay experiments to investigate a harmful algal bloom in the Bay of Seine (French-English Channel) *Marine Pollution Bulletin* **168** 112387
[41] Finkel Z V, Irwin A J and Schofield O 2004 Resource limitation alters the 3/4 size scaling of metabolic rates in phytoplankton *Mar. Ecol. Prog. Ser.* **273** 269-79
[42] de Vasconcelos E, Vasconcelos J B, Reis T N D, Cocentino A D M, Mallea A J A, Martins G M, Neto A I and Fujii M T 2019 Macroalgal responses to coastal urbanization: relative abundance of indicator species *J. Appl. Phycol.* **31** 893-903

IOP Conf. Series: Materials Science and Engineering 1245 (2022) 012012 doi:10.1088/1757-899X/1245/1/012012

Solent oyster enhancement reef concrete colonisation trials

K J Collins, C Hauton, A C Jensen and J J Mallinson

Ocean and Earth Science, University of Southampton, SO14 3ZH, UK

Abstract. The Interreg funded project RECIF (2012-2015) developed waste shell concrete formulations specifically for artificial reefs. Its successor MARINEFF continues these evaluating further formulations to maximise both physical performance and endurance with biological attractiveness. Four formulations were produced by ESITC Caen for trials based on combinations of two cement types (CEM II and CEM IV) with either coarse aggregate or crushed oyster shells replacing 20% of the aggregate. The most extensive colonisation trials were carried out with 5x5x2cm blocks by both UK and French partners. These test blocks also had smooth and rough faces. This paper describes the results of Southampton biological colonisation trials. Four blocks, one of each formulation along with an oyster shell cut to the same area were secured in a framework and pairs of these were suspended vertically at 6 fully saline sites around the Solent region (Calshot, Beaulieu River, Hamble River, Itchen River, Portsmouth and Langstone Harbours) from May to August 2019. On recovery these were analysed for Total Biomass, Algae (PAM & %cover), Anemones (n), Ascidians (%cover), Barnacles (n & % cover), Bryozoans (%cover), Hydroids (%cover), *Mytilus edulis* (n), *Pomatoceros sp.* (n), Sponges (%), *Spirorbis sp* (n). There were considerable differences in colonisation between the sites which are described in detail, but overall in terms of overall biomass the formulation using CEM V and 20% oyster shell performed as well as the oyster shell. Given that it is accepted that oysters are attracted to natural shell substrate, this concrete was promising for the choice mix of the final oyster reef modules deployed at Calshot, September 2020.

1. Introduction

The Interreg funded project RECIF (2012-2015) [1] developed waste shell concrete formulations specifically for artificial reefs. Its successor MARINEFF (2018-2022) continued the development of these by evaluating further formulations to maximise both physical performance and endurance with biological attractiveness. Both projects were selected under the European Cross-border Cooperation Programme INTERREG IV A between France (Channel) - England, co-funded by the ERDF. The project bringing together French and British partners and is supported by the Regional Authorities in Basse-Normandy. Beyond the Channel, in Spain the most important and biggest waste generator of waste shell is from Galicia's bivalve farming industry there is a particular interest in their reuse [2, 3].

The authors have long experience with assessing the environmental computability of novel materials; cement stabilised pulverised fuel ash [4-8] and waste tyres [9-11]alongside conventional concrete as the control material. The Poole Bay artificial reef was originally constructed in 1989 deploying 8 mounds of blocks 5m across by 1m high, on a 10m grid in 12m depth water. In 1998 the reef area was doubled in size adding tyre and concrete units. The use of both the above waste materials was potentially controversial and a key component of the monitoring programme was analysis of the organisms growing on the reef units to determine if compounds in the wastes had transferred across. No evidence of chemical contamination was found in either case but another approach was also employed, monitoring the growth of organism, percentage cover of different taxonomic groups and biomass against that on conventional concrete controls deployed at the same time. The logic of this approach was that even if there were no detectable chemical contamination then the effect of the material could potentially be more subtle influencing settlement and growth rates influencing the developing colonising ecosystem. Shells are a natural, biogenic material for which there is ample evidence of excellent colonisation [12] [13] and ordinary concrete is accepted as a suitable material for marine construction including artificial reefs, thus no chemical contaminant studies were included in both the RECIF and MARINEFF shell-concrete trials. Both programmes compared colonisation of shell concrete with ordinary non-shell concrete and the effect of surface texture on the growth of colonisers and included extensive studies of the physical performance of the formulations by ESITC Caen [14,15] and their mineralogical composition by University of Exeter [16].

Beyond refining shell concrete, the goal of the MARINEFF programme is to use the formulations to enhance and protect coastal and transitional water ecosystems in cross-border Channel regions. Today, marine infrastructures (such as breakwaters, quays, piers, groynes, and moorings) do not incorporate the enhancement of coastal and transitional water ecosystems. The project aims to realise new biomimetic marine infrastructures to improve the ecological status of coastal and transitional Channel waters, as well as to involve professionals and stakeholders in the project.

Globally, an estimated 85% of oyster reefs have been lost, making oyster reefs one of the world's most imperilled marine habitats. The native oyster is now functionally extinct in many areas around Europe, having declined by over 90% [17].The European flat oyster, *Ostrea edulis*, is an important ecosystem engineer that has been progressively disappearing from European coasts over the last century mainly due to overexploitation, habitat degradation and disease. It is now the subject of many conservation and restoration programs throughout Europe [18]. The Southampton focus is on supporting the current Solent Oyster Restoration Project [19,20] with specially designed oyster settlement structures recently deployed in the Solent. Settlement of *Ostrea edulis* is determined by the availability of hard substrata [21]. A pilot study suggested that the culture of oysters on elevated reef structure represents at least a partial solution to improve *Ostrea edulis* physiology for restoration [22].

SIFCA (2018), working with the Blue Marine Foundation [23] has declared the following objectives for Solent oysters:

- Increase oyster populations to levels that restore important ecological functions, including water filtration and nutrient cycling; habitat and community structure; and adequate brood stock to sustain regional populations.
- Achieve a sustainable oyster fishery and support jobs in the fishery sector through the introduction of effective systems of managing harvest

2. Methods

In April 2019 ESITC Caen, University of Caen, France produced test blocks of four candidate mixes:

- M1: Standard Concrete with Cement CEM V+ 0% of Shells
- M2: Concrete with Cement CEM V + [40 to 60%] Shells
- M3: Standard Concrete with Cement CEM II + 0% of Shells
- M4: Concrete with Cement CEM II + [40 to 60%] Shells

Small (5 x 5 x 3 cm) test blocks were produced with a smooth and rough side. Biological colonisation of these small blocks was studied by both by the University of Southampton and University of Caen marine station of Luc Sur Mer. At the beginning of May 2019 Southampton deployed vertical racks of each of the 4 mixtures plus an oyster shell cut to the same size as the blocks (Figure 1) at 7 sites reflecting the estuarine environment of the region (Figure 2):

- Calshot and Beaulieu, 4 replicate racks at each site, attached to wooden piles below low-water.
- Hamble, Itchen, Portsmouth, Langstone, and Chichester, 2 replicates at each site, suspended in oyster cages below floating pontoons, potentially exposing them to planktonic oyster larvae.

In parallel to the above, following a range of immersion periods for test blocks physical properties were determined by ESITC Caen, University of Caen and mineralogical composition of the mixes were analysed by the University of Exeter

After 4 months immersion the 18 Southampton test racks (4 concrete mixes + oyster shell) the rough and smooth sides were analysed for percentage cover (%) of organisms and where appropriate the number (n) of individuals: algae (%), anemone (n), ascidians (%), barnacles (n & %), bryozoans (%), hydroids (%), *Mytilus* (n), *Pomatoceros* (n), sponges (%), *Spirorbis* (n). The degree of algal colonisation was determined by Pulse Amplitude Modulation (PAM) fluorescence measurements (10 replicates) using a Diving-PAM (Heinz Walz GmbH). Total biomass (dry weight) of each test block/shell was measured.

Figure 1. Test racks of 5x5cm blocks and single 15cm cubes prior to deployment at Calshot showing the demonstration site, the former power station cooling water outfall tower in the background.

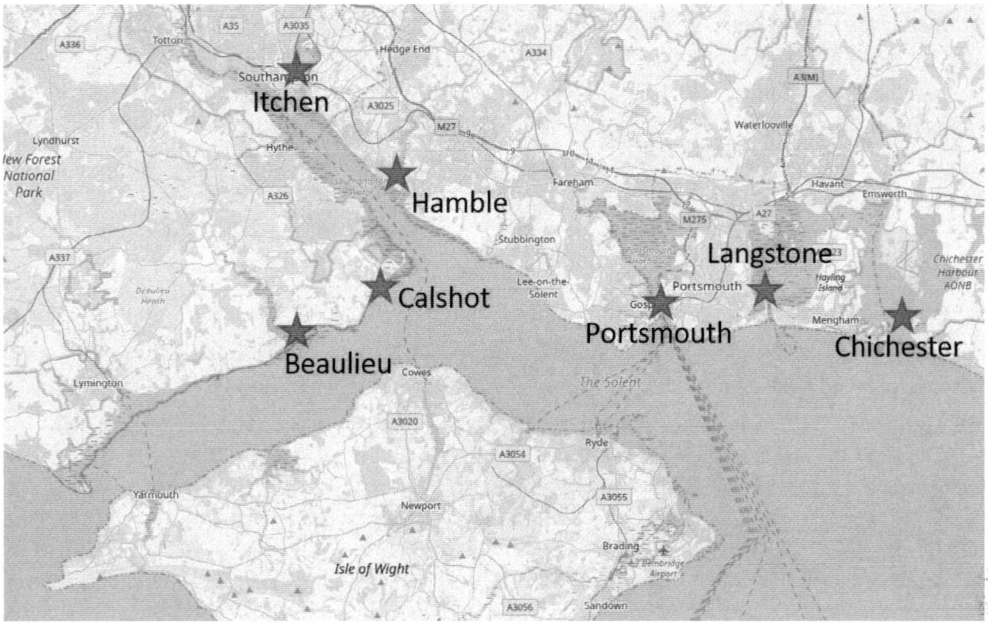

Figure 2. Locations of the deployment sites for the test racks.

3. Results

At the Hamble, Itchen, Portsmouth, Langstone, and Chichester, the racks suspended in cages with mature oysters below floating pontoons were potentially exposed them to planktonic oyster larvae released May-June each year. No settled spat were detected on the test surfaces, but since the larvae have pelagic planktonic stage of ~10 days then in the tidal estuary they are rapidly swept away before they can settle.

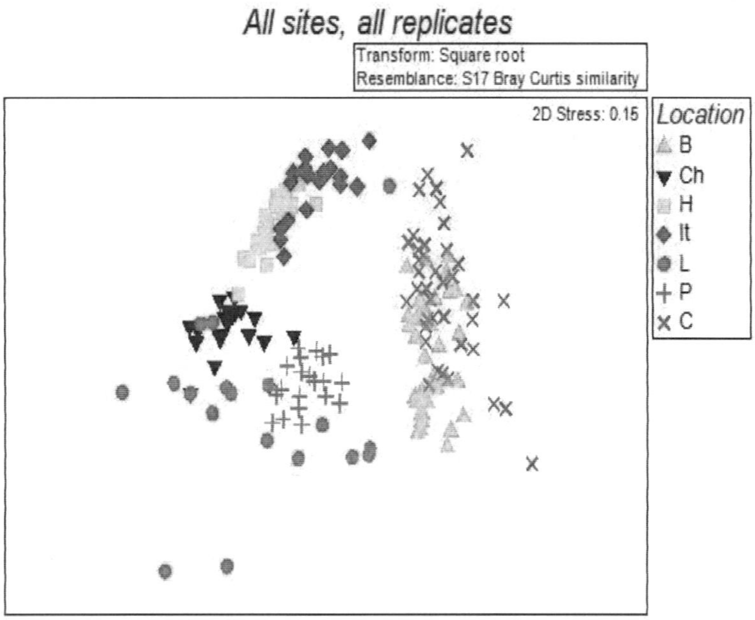

Figure 3. MDS analysis of combined colonisation data from all surfaces and replicates by site (see Figure 2)

Non parametric analysis using PRIMER [23] of the quantitative colonisation was undertaken. One way ANOSIM analysis by location showed a significant difference in community (grouping all mixes and surfaces between all locations (R=0.767, P=0.001). Figure 3 shows 3 distinct groupings reflecting different water bodies: Solent = Beaulieu and Calshot, Southampton Water = Itchen and Hamble, shallow eastern harbours = Portsmouth, Langstone and Chichester. Beaulieu was different from, Chichester (R=0.998, R=0.001), Hamble (0.996, p=0.001) Itchen (R=0.957, p=0.001) Langstone (R=0.91, p=0.001), and Portsmouth (0.879, p=0.001) but similar to, but not statistically the same as Calshot (R=0.34, p=0.001).

ANOSIM analysis of all mixes (including oyster shell) across all locations showed no significant difference (R = -0.002, P=0.554). This suggests that location is much more important in community development than mix or surface. Examining the 2 sites with most replicates ANOSIM showed no differences in community between mixes or surfaces: Beaulieu R=0.153, P=0.006 and Calshot R=0.064, p=0.10. Figure 4 suggests the biomass was higher, but not significantly so, on mix 4 with shell and oyster shell alone.

Comparison of the colonisation of the rough vs smooth sides of the test blocks did not show a statistically significant difference (Mann-Whitney Rank Sum Test, n= 695, P = 0.480).

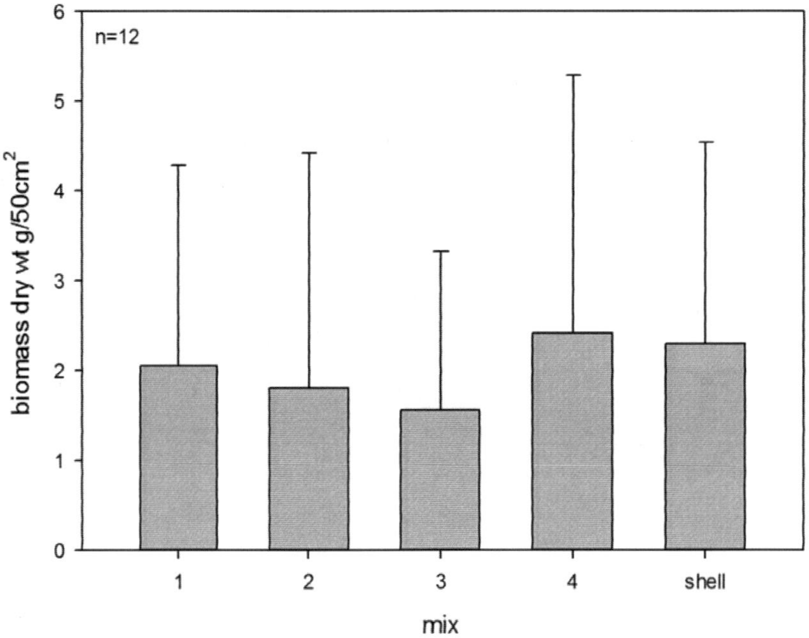

Figure 4. Average biomass (and standard deviation n=36) of the colonisation across rough and smooth sides of the 4 concrete test mixes plus oyster shell.

IOP Conf. Series: Materials Science and Engineering 1245 (2022) 012012 doi:10.1088/1757-899X/1245/1/012012

Table 1. Abundance of species found on all test blocks at the 7 tests sites (Figure 2): r = rare, o = occasional, f = frequent, c = common, a = abundant, p = present, ?=ID uncertain

Phylum	group	Species	B	C	I	H	P	L	Ch
PORIFERA	sponges	*?Amphilectus fucorum*			p				
		Halichondria sp.		r					
		Sycon ciliatum				o	f	p	p
CNIDARIA	hydroids	*Plumularia setacea*	o					f	
	anemones	*Diadumene lineata*			o				
ANNELIDA	worms	*Ficopomatus enigmaticus*			f				
		Sabella sp.	r						
		Spirobranchus sp.	f	r	o	o	p	p	p
		Spirorbis sp.	c	r		o	p	p	
CHELICERATA	pycnogonids	*Ammothea hilgendorfi*	f						
CRUSTACEA	barnacles	*Austominius modestus*	o	r	p	f	p		p
		Balanus crenatus			c	o			
	caprellid	*Caprella ?linearis*	o			f			
	amphipod	*Amphipoda indet.*	p		p				
	isopod	*Cymodoca truncata*	f						
	prawn	*Hippolyte varians*	r						
	crabs	*Pilumnus hirtellus*	f						
		Pisidia longicornis	f						
MOLLUSCA	sea slug	*Thecacera pennigera*	o						
	bivalves	*Mytilus edulis*			o				
BRYOZOA	foliose	*Bicellariella ciliata*	o						
		Bugula neritina				p	p		f
		Bugulina fulva		r					
		Crisularia plumosa	p						
		Scrupocellaria scruposa	p						
		Tricellaria inopinata	p	r	p	p		o	o
	encrusting	encrusting bryozoan on *Mytilus*			p				
		Watersipora cucullarta				p		r	
ECHINODERMS	brittlestars	*Amphipholis squamata*				r			
TUNICATA	seasquirts	*Ascidiella aspersa*	f				p	p	r
		Ciona intestinalis	o				p		r
		Dendrodoa grossularia			r				
		Molgula sp.				r			
	encrusting	*Botrylloides diegensis*	o						
		Botrylloides ?leachii	o		p	p			
		Botrylloides violaceum	o		p	f		p	
		Botrylloides sp.	o					p	p
		Botryllus schlosseri	o				p	r	p
	didemnids	*Didemnum/Lissoclinum/Diplosoma*	o						p
		Diplosoma spongiforme	r						
ALGAE	green	*Bryopsis plumosa*		r					
		Chaetomorpha linum	r						
	brown	*Ectocarpus aggregation*		r					
		Fucus sp.							
	red	*? Metacallophyllis laciniata*		r					
		Ceramium sp.		r					
		Halurus flosculosus		r					
		Polysiphonia sp.		r					
		filamentous red						o	

4. Discussion

The recent MARINEFF shell-concrete colonisation trials confirm those from earlier RECIF programme, demonstrated rapid growth of epifauna and epiflora at least equivalent to that on the control materials. Whilst to date there has been an indication that there has been greater biomass growth on the shell-concrete compared to ordinary, non-shell concrete this is not statistically significant.

A key aim of the current programme was to produce oyster enhancement reef modules focussing attention on oyster settlement. Shells are a biogenic material and would be expected to be readily colonised. An investigation into the colonisation of *Chlamys opercularis*, the Queen Scallop, shells in the Isle of Man found that younger scallops were predominantly colonised by sponges whilst the shells of older scallops showed a greater variety of encrusting organisms including many bryozoans [12]. Oyster larvae respond to chemical cues to induce settlement behaviour. Such cues occur both naturally in the environment and from oyster tissue [24], consequently oyster larvae often settle on conspecifics allowing reefs to develop over time. The colonisation of oyster shells, both *Ostrea edulis* and *Crassostrea gigas* was studied in Poole Bay and Poole Harbour with *Ostea edulis* spat recorded at both sites [13] but finding a greater abundance occurring on elevated reef structures compared to the seabed [22]

The surface texture of the shell-concrete appears to influence colonisation biomass. The greater mass on smooth shell-concrete is surprising given that there is often no selection between smooth and rough surfaces [26]. *Ostrea edulis* larval recruitment was found to be influenced more by substrate surface texture and micro-topography than by substrate formulation within the ranges tested [27]. The best formulation of the nine formulations tested was that containing the highest proportion of shell (50%), although this trend was not significant. The RECIF programme found that the concrete strength decreased with increasing shell inclusion [1] concluding with a maximum of 40%. The current MARINEFF programme increased this maximum to 60% using different cements. The test blocks used in this study were made in open moulds, the base producing the smooth surface and the open upper surface left irregular. However this still left a film of concrete over the aggregate and crushed shell filler which explains the lack of significant difference in colonisation between mixes found in this study. The refinement during manufacture of the subsequent demonstration oyster modules was to pressure wash the interior surfaces before the concrete had fully set, exposing clean shell fragments on the surface and providing a much more irregular surface. Settlement studies of oyster larvae on surfaces at different angles found most on the undersides of horizontal and sloping surfaces [28]. This was similarly reported during the RECIF trials [13] which led to the adoption and manufacture (by MARINEFF partner TPC, Cherbourg, France) of the hollow prism modules subsequently deployed off Calshot in September 2020.

Figure 5. An oyster module prior to deployment, showing the pressure washed interior surface with exposed shell.

Acknowledgements
This study was supported by the MARINEFF project, selected by the European cross-border Interreg IVA France (Channel)-England programme co-funded by the ERDF. Luke Helmer, Blue Marine Foundation kindly deployed the test racks in oyster cages. Alex Buckley and Antoni Ozturk were employed through MARINEFF funded internships to support the colonisation trials.

References

[1] Boutouil, M. and S. Le Boulanger. *RECIF Conference on artificial reefs: from materials to ecosystem.* in *RECIF Conference on artificial reefs: from materials to ecosystem.* 2015. ESITC Caen, France.
[2] Carral, L., et al., *Social interest in developing a green modular artificial reef structure in concrete for the ecosystems of the Galician rias.* Journal of Cleaner Production, 2018. **172**: p. 1881-1898.
[3] Carral, L., et al., *Assessment of the Materials Employed in Green Artificial Reefs for the Galician Estuaries in Terms of Circular Economy.* International Journal of Environmental Research and Public Health, 2020. **17**(23).
[4] Collins, K.J., A.C. Jensen, and A.P.M. Lockwood, *Fishery Enhancement Reef Building Exercise.* Chemistry and Ecology, 1990. **4**: p. 179-187.

[5] Collins, K.J., A.C. Jensen, and A.P.M. Lockwood, *Artificial Reefs: using coal fired power station wastes constructively for fishery enhancement.* Oceanologica Acta, 1991. **11**: p. 225-229.

[6] Collins, K.J., A.C. Jensen, and A.P.M. Lockwood, *Stability of a coal waste artificial reef.* . Chemistry and Ecology, 1992. **6**(73-93).

[7] Jensen, A.C., et al., *Lobster (Homarus gammarus) movement on an artificial reef; the potential use of artificial reefs for stock enhancement.* . Crustaceana, 1994. **67**(2): p. 198-211.

[8] Collins, K.J., et al., *Evaluation of stabilised coal-fired power station waste for artificial reef construction.* Bulletin of Marine Science 1994. **55**(2-3): p. 1265-1278.

[9] Collins, K.J., A.C. Jensen, and S. Albert, *A review of waste tyre utilisation in the marine environment.* Chemistry and Ecology, 1995. **10**: p. 205-216.

[10] Collins, K.J., et al., *Scrap tyres for marine construction: environmental impact.* , in *Recycling and Reuse of Used Tyres*, R.K. Dhir, M.C. Limbachiyya, and K.A. Paine, Editors. 2001, Thomas Telford: London. p. 149-162.

[11] Collins, K.J., D. Howell, and T.R. Fisher, *Tyres in coastal and flood defence engineering – environmental monitoring.*, in *Sustainable Waste Management*, M.C. Limbachiyya, Editor. 2004, Thomas Telford: London. p. 227-234.

[12] Ward, M. and J.P. Thorpe, *Distribution of encrusting bryozoans and other epifauna on the subtidal bivalve Chlamys opecularis.* Marine Biology, 1991. **110**: p. 253-259.

[13] Deane, S., et al., *Epifaunal colonisation on shells of native European oysters, Ostrea edulis, and non-native Pacific oysters, Crassostrea gigas.*, in *RECIF Conference on artificial reefs: from materials to ecosystem*, M. Boutouil and S. Le Boulanger, Editors. 2015, ESITC Caen, France.

[14] Cuadrado, H., et al., *Physical properties and mechanical behaviour of concrete made with crushed queen scallop shells.* , in *International symposium on environmentally friendly concrete - ECO-Crete.* 2014: Reykjavik, Iceland. p. 21-28.

[15] Cuadrado, H., et al., *Properties of ordinary concretes incorporating crushed queen scallop shells. Materials and structures., 49: 1 – 12, 2015*

[16] Sawyer, T.T., et al., *Mineralogical properties of marine shell products for the construction of concrete artificial reefs*, in *RECIF Conference on artificial reefs: from materials to ecosystem*, M. Boutouil and S. Le Boulanger, Editors. 2015, ESITC Caen, France. p. 19-27.

[17] Preston, J., et al., *European Native Oyster Habitat Restoration Handbook.* 2020: p. 68.

[18] NORA. *Native Oyster Restoration Alliance – Europe*. 2021.

[19] Harding, S., L. Nelson, and T. Glover, *Solent Oyster Restoration Project Management Plan.* 2017, Blue Marine Foundation. p. 47.

[20] SIFCA, *Solent Oyster Management Plan.* 2018, Southern Inshore Fisheries and Conservation Authority. p. 16.

[21] Smyth, D., et al., *Settlement of Ostrea edulis is determined by the availability of hard substrata rather than by its nature: Implications for stock recovery and restoration of the European oyster.* Aquatic Conservation-Marine and Freshwater Ecosystems, 2018. **28**(3): p. 662-671.

[22] Sawusdee, A., et al., *Improvements in the physiological performance of European flat oysters Ostrea edulis (Linnaeus, 1758) cultured on elevated reef structures: Implications for oyster restoration.* Aquaculture, 2015. **444**: p. 41-48.

[23] Blue. *Solent Oyster Restoration project*. 2015

[24] Clarke, K.R. and R.N. Gorely, *PRIMER v6: User Manual/Tutorial.* 2006, Plymouth Marine Laboratory: PRIMER-E Ltd. 190.

[25] Bonar, D.B., et al., *Control of oyster settlement and metamorphosis by endogenous and exogenous chemical cues.* Bulletin of Marine Science, 1990. **46**(2): p. 484-498.

[26] Crisp, D.J. and J.S. Ryland, *Influence of filming and of surface texture on the settlement of marine organisms.* Nature, 1960. **185**((4706)): p. 119-119.

[27] Potet, M., et al., *Which concrete substrate suits you? Ostrea edulis larval preferences and implications for shellfish restoration in Europe.* Ecological Engineering, 2021. **162**.

[28] Cole, H.A. and E.W.K. Jones, *Some Observations and Experiments on the Setting Behaviour of Larvae of Ostrea edulis.* ICES Journal of Marine Science, 1939. **14**(1): p. 86-105.

Mineralogical and microtextural characterization of eco-friendly concrete for marine infrastructures: Methodologies for the *MARINEFF* Project

V Ramos, J C Andersen and G K Rollinson

Camborne School of Mines, University of Exeter, Penryn Campus, Treliever Road, Penryn, Cornwall, TR10 9FE, UK

Abstract. The European cross-border Project *MARINEFF* explores the use of eco-friendly concrete materials in the construction of marine structures (oyster bed modules, dykes, boat moorings and rockpools). The structures and materials are designed to protect and improve the local biodiversity of the marine environment of the Channel / La Manche. This paper focuses on the mineralogical and microtextural characterization of the concrete materials. The aim is to introduce how advanced analytical methods have been employed to gain an understanding of the evolution of the materials, in particular their response to immersion into the sea. Our methods include the traditional techniques for examining concrete materials (optical microscopy, scanning electron microscopy, X-ray diffraction), but also introduces the benefits of mineralogical methods that go beyond the current "state-of-the-art" (automated scanning electron microscopy by QEMSCAN®, electron-probe microanalysis). Our results offer detailed insight into the controls on the compositional and structural evolution of the materials.

1. Introduction

Project MARINEFF (MARine INfrastructures EFFects) arises from the need to protect the ecological status of coastal waters at a time when climate change, fishing, recreational use and coastal development are significantly affecting the coastal dynamics and marine biodiversity. The project is funded by the European Regional Development Fund, coordinated by the École Supérieure d'Ingénieurs des Travaux de la Construction de Caen (ESITC), France, and involves several partner institutions across England and France. The project aims to develop novel eco-friendly concrete materials that are affordable, durable and easy to incorporate into marine infrastructure projects. The ultimate goal is to enhance and protect costal and transitional water ecosystems in the cross-border Channel region. The project also explores how the integration of habitats directly into the marine infrastructures may lead to increasing biodiversity. The other main innovation aspect of the project is the use of recycled oyster shell waste into the concrete as a partial replacement of aggregate (up to 20%). In recent years, the incorporation of seashell waste in concrete and mortar, whether as aggregate replacement or as a filler, has gained significant attention as a route for the industry to meet the increasing sustainability challenges in the marine construction field (e.g. [1-6]).

The concrete infrastructures developed in the *MARINEFF* project include oyster beds, dykes, underwater boat moorings and artificial rockpools. The structures were deployed at test sites in the coastal waters of England and France, and they are being monitored periodically for mineralogical and microtextural changes as well as biocolonisation.

Raw materials as well as concrete mixtures were investigated at Camborne School of Mines, University of Exeter, UK to determine their mineralogical and microstructural properties prior to and after immersion. Our work aims to develop an understanding of how the concrete materials respond to exposure in the marine environment. Seawater attack on concrete is caused primarily by reactions of magnesium and sulphate with the cement paste phases, such as calcium hydroxide (CH), aluminate phases and calcium silicate hydrates (CSH). The reactions lead to the formation of brucite, magnesium silicate hydrates, gypsum and ettringite, all of which cause deterioration (e.g.[7-9]). This deterioration accelerates water and chloride ingress into the concrete, which may lead to the rapid corrosion of reinforced structures.

In this paper, we present the research methods for the mineralogical and microtextural investigation of concrete formulations from the *MARINEFF* project. This summary aims to demonstrate the benefits of these methodologies in understanding and interpreting the behaviour of concrete in the marine environment. It is our hope, that a wider appreciation of these tools can advance knowledge of concrete degradation through the cooperation between mineralogists and geoscientists, civil engineers, biologists, ecologists and the construction industry.

2. Methodologies for the mineralogical and microtextural characterization of marine eco-engineering concrete

A successful investigation of concrete (as any other materials) should always start with a detailed understanding of the variability within the material as well as the interactions that it has with the environment. This understanding will dictate where and how samples are selected for study, how many samples are required, and what sizes they need to be. Samples may be collected to be representative of the bulk of the material of interest, or they may be targeted to specific parts to allow for the detailed exploration of an area of interest. Detailed considerations on concrete sampling can be found, for example, in [10-12].

A wide range of techniques are currently available for the investigation of hardened concrete. Most of the established methods relate to bulk testing of physical or chemical characteristics, or they are based on optical microscopy, which is highly subjective. A key role for the geoscientist or materials scientist is to select the most adequate and effective methods to provide the information that is required.

For the mineralogical and microtextural investigation of concrete, it is important to collect information in the field – particularly to document environmental controls and variability. However, the more detailed observations require laboratory study. The examination will usually progress from direct observation and optical microscopy to more advanced (and costly) analytical methods. In the following sections, we introduce and summarize the methodologies used for investigating the mineralogy and microtextures of concrete materials in the *MARINEFF* project. Our examinations partly follow established routines, but they also give examples of how advanced analytical techniques that are not routinely employed to study concrete offer additional insight into the materials.

2.1. Optical petrography

Optical petrography is the first step to investigate concrete, mortar and related components. This technique gives essential insight into the mineralogy, compositional features, quality, workmanship and conformity with standards and specifications. Petrography is also used to identify concrete degradation mechanisms such as alkali-aggregate reactions, sulphate attack, seawater attack, freezing and thawing, fire damage, among others. A careful petrographic examination of representative samples can by itself reduce or even avoid the need to use other more expensive investigation methods [7].

A good petrographic examination starts with the macroscopic examination of the concrete samples, which can be complemented by the use of a hand lens and/or a stereomicroscope. This initial assessment can provide information regarding the appearance and condition of the cement paste; the distribution, granulometry and type of aggregates; voids and cracks (and their infilling); the carbonation depth; the presence and position of reinforcement and the extent of associated corrosion; and the presence of any superficial evidence of degradation of the concrete [11-13].

A detailed petrographic examination requires the concrete to be prepared into a polished thin section (a 30μm thick slice of material mounted on a glass slide) that allows for examination with a polarizing microscope under plane- and cross-polarized transmitted light as well as the scanning electron microscope (SEM). Descriptions of concrete thin section preparation can be found, for example, in [7,13-14].

The examination of concrete under the polarizing microscope allows for detailed evaluations of the structure and mineral abundances in the aggregate and binder. Observations give further detail on voids, micro-fracturing and other structural defects, as well as evidence for the progression of deleterious reactions. Particular aspects of study include the identification of the types of aggregates; the size, distribution and abundances of phases in the binder; the presence of supplementary cementitious materials (SCMs) such as blast furnace slag (GGBS), fly ash (FA) or silica fume. Interpretations can also be made of the type of cement clinker; the presence of high alumina cement; products (and causes) of cement paste deterioration; and the original total water to cement ratio. The optical microscope, however, does not give direct information on the chemical composition of concrete. It is also not an adequate tool for the investigation of chloride ingress or the presence of plasticisers.

Currently, optical microscopes usually have digital cameras that allow for the acquisition of high-resolution colour photomicrographs to document critical observations. Image processing techniques can be employed for component identification as well as the quantification of fractures and voids [7,11]. The ASTM C856/C856M standard [15] specifies the method of petrographic examination of hardened concrete. Petrographic study of the *MARINEFF* samples were carried out at Camborne School of Mines using a Nikon Eclipse E600 Pol optical microscope with a Nikon Digital Sight 5MP camera that captures images using the NIS Elements F3.0 software (Figure 1a). The optical petrography focused on the identification of the main mineral phases with special emphasis on minerals formed during seawater attack, the presence of carbonated cement paste, the effectiveness of the aggregate/seashell-cement paste bonding, the presence of cracking, and the shape and distribution of voids and presence of infillings. Figures 1(b) to (f) illustrate some of the observed aspects of the investigated concrete under the optical microscope.

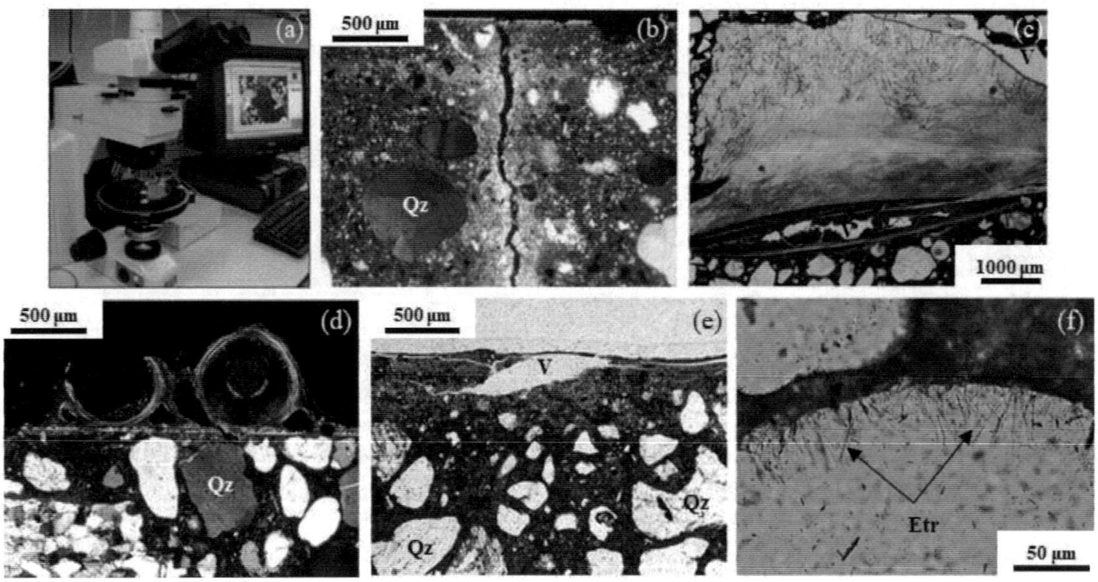

Figure 1. (a) Nikon Eclipse E600 Pol microscope coupled with a digital camera; (b) Carbonated cement paste (lighter colours) at the margins of an open drying-shrinkage crack. An aragonite crust partially covers the concrete block; (c) Several voids in the spongy layer of an oyster shell. In the upper right corner it is evident the lack of bonding with the cement paste (d) Calcareous organisms settled on the top of a concrete block protecting it from erosion; (e) Presence of microcracks at the edge of a concrete block resultant from seawater attack; (f) Ettringite acicular crystals from sulphate attack on the cement paste lining an air void. Concrete blocks in figures (b) and (c) were immersed for 3 and 6

months, respectively. Concrete blocks in figures (d), (e) and (f) were immersed for 12 months. Qz – quartz; V – void; Etr – ettringite.

2.2. Scanning electron microscopy and electron probe micro-analysis

The scanning electron microscope (SEM) in general and depending on the model used, offers image magnifications up to more than 50.000x and resolutions down to 1 nm. In these respects, they supplement the optical microscopy by allowing for observations of much smaller structures and much finer materials – such as concrete binder. The electron beam interactions with the specimen generate X-rays, which allow for the collection of some chemical compositional information via an energy dispersive detector (EDS). Electron imaging involves the collection of secondary electrons (SE) and backscattered electrons (BSE). Secondary electrons imaging is most reliable for the study of the topography of the sample, whilst BSE images can show contrast due to variations in chemical composition of a specimen. The generation of characteristic X-rays facilitate in situ chemical analysis and identification of minerals or phases by the interpretation of these spectra [16-18]. The electron-probe micro-analyser (EPMA) is optimised for quantitative chemical analysis but has a lower spatial resolution. The EPMA works at a higher beam current, that enhance X-ray signals and minimise noise. The EPMA is fitted with wavelength dispersive spectrometers (WDS) that have higher sensitivity and better spectral resolution than the SEM-EDS [18]. The signals are calibrated against standard reference materials for quantitative analysis. The analysis allows for detailed determination of mineral compositions and formulae (e.g. [19-21]). The SEM-EDS and EPMA require samples to be coated with a 25nm carbon film to avoid the build up of electrical charge on the surface [16].

The *MARINEFF* samples were examined with a Tescan Vega 3 SEM-EDS system (Figure 2(a)) and the JEOL JXA-8200 Superprobe (Figure 2(b)). The SEM-EDS is particularly useful to confirm and identify the presence of concrete phases and seawater chemical attack reaction phases, such as ettringite, thaumasite and brucite, and also to observe microtextures and fractures at the nanometre scale. Figures 2(c) to (f) correspond to BSE images of several aspects of the mineralogy and microtexture present in the concrete samples investigated in the *MARINEFF* project.

The EPMA is particularly useful to explore chemical substitution in minerals. It can also be used for X-ray elemental mapping at micron-scale, which aids in understanding the spatial distribution of deleterious substances in the cement paste, around fractures, or on aggregate surfaces (e.g. determination of the penetration depths for chloride and sulphate ions).

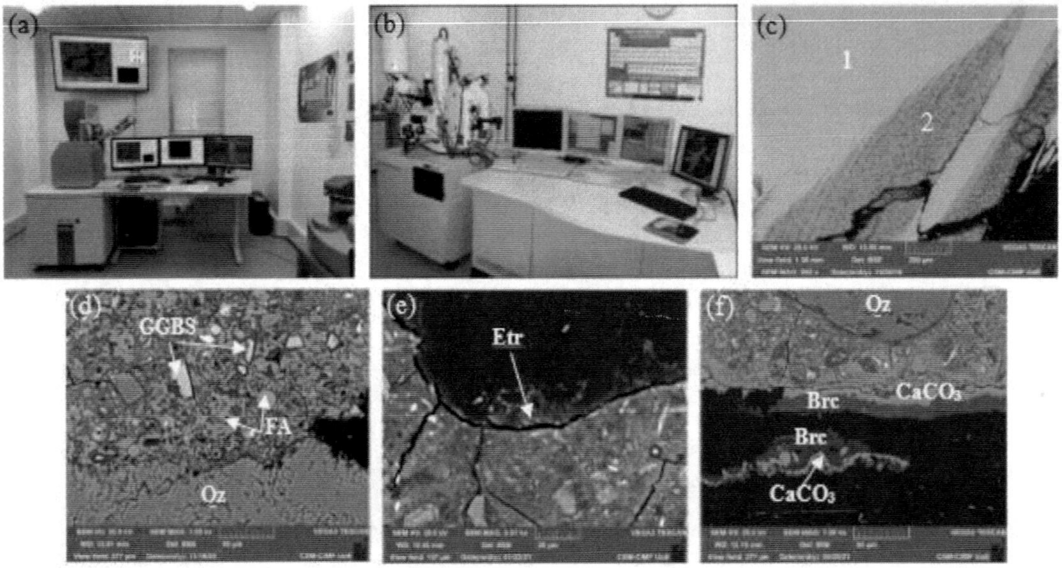

Figure 2. (a) SEM Tescan Vega 3 GMU; (b) JEOL JXA-8200 Superprobe; (c) Oyster shell – characteristic textures of the prismatic layer (1) and of the porous bulky layer (2); (d) Presence of GGBS and FA in concrete formulated with CEMV; (e) Acicular crystals of ettringite lining an air void in a concrete sample immersed for 3 months. The microcracks are resultant from the processes involved in the preparation of the polished thin section; (f) Presence of a cracked brucite layer associated with $CaCO_3$ at the edge of concrete immersed for 12 months.

2.3. Automated scanning electron microscopy analysis

The QEMSCAN® is a SEM-EDS system that has been optimised for the rapid analysis of mineralogical materials to provide information on mineral distributions in samples [22]. The instrument combines BSE and characteristic X-rays [23-26] and compares them to a customisable database of reference minerals. The QEMSCAN® allows for the full identification and quantification of the minerals present on a sample surface as well as of particle size distributions and systematic relationships between the different mineral phases. The scan of a polished surface also allows a false colour image or map to be created (in Fieldscan mode) at a micron-scale corresponding to a chemical image of the mineralogy and requires interpretation for mineral identification [27].

The QEMSCAN® has the advantage of being able to identify and quantify minerals that are not obvious from the optical microscopy. Optical observation is invariably subjective and dependent on the skills of the observer. However, similar to optical microscopy, particle size data suffer from the problem of stereology (2D particle data representing a 3D particle) underestimating true sizes [22,25,27,28]. A further issue is that touching particles of the same mineral are not distinguishable (e.g. quartz touching quartz gives a single grain) [26].

The QEMSCAN® 4300 at Camborne School of Mines is built around a Zeiss EVO 50 SEM with 4 x Bruker SDD EDS detectors. While the sample measurement is automated, the data processing requires manual evaluation in order to ensure that results are correct (quality control). Sample measurement uses iMeasure version 4.2SR1 software for data collection at the default 1000 X-ray counts per point and 25 kV, and iDiscover 4.2SR1 and 4.3 software for data processing.

For the *MARINEFF* concrete samples (Figure 3(a)), the QEMSCAN® analysis has been employed to identify the shape, size and distribution of aggregates/seashells, voids and porosity. It has helped to visualise the effectiveness of aggregate/seashell bonding with the binder; the presence, extension and width of cracks and the potential existence of infillings; the extent of carbonation; and the identification of secondary products resultant from seawater attack.

Figures 3(b) and (c) present false colour mineral maps of examples of concrete polished thin-sections. The observation of these maps allow a full and intuitive identification of the concrete mineralogy and its relations besides providing information related to the size, shape and distribution of aggregates, seashells and porosity, but also the extent of cracking. The concrete in Figure 3(b) corresponds to a non-immersed sample formulated with CEMII, quartz-rich aggregates and oyster shells mainly composed by calcite. The main component of the cement paste is CSH. The false colour map in Figure 3(c) corresponds to a scan of the edge of a concrete block made with CEMV and quartz-rich aggregates and immersed for 3 months. The presence of a Mg silicate-rich zone followed by a Ca sulphate-rich zone is a result of seawater attack. It is notable that these zones are deflected around a drying-shrinkage crack that allows the further penetration of the reaction into concrete. In this case, the cement paste is mainly composed by calcium alumina silicate hydrates (CASH) that have been carbonated in the outermost zone. Internal carbonation is developed to a lesser extent (lighter areas), as it can be observed in Figure 3(d).

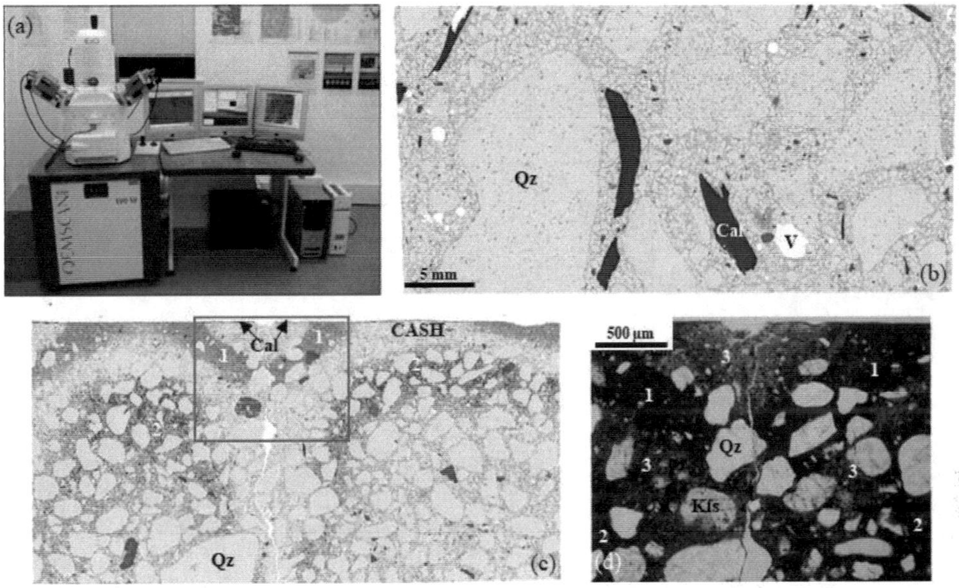

Figure 3. (a) QEMSCAN® 4300; (b) False colour mineral map of non-immersed concrete made with CEMII, quartz-rich aggregates and oyster shells. The cement paste is dominated by CSH. (c) False colour map of the edge of a concrete block immersed for 3 months and formulated with CEMV; (d) Photomicrograph in plane polarized light from the insert in (c). Qz – quartz; Cal – calcite; V – void; CASH – calcium alumina silicate hydrate; Kfs – K-feldspar; 1 – Mg silicate-rich zone; 2 – Ca sulphate-rich zone; 3 – carbonated cement paste.

2.4. X-ray diffraction analysis

X-ray diffraction (XRD) is a standard bulk method used for the direct identification of crystalline materials and minerals, in this case using an homogenised powdered sample. While it can be used for the quantification of minerals, it lacks the aspect of spatial distribution that can be gained from the QEMSCAN®. Also, amorphous phases which are common in concrete do not generate unique diffraction patterns, although their presence can be deduced by the background intensity maxima [29]. Sample preparation for XRD analysis requires the powdering of the material (<50 µm) in order to achieve a homogeneous sample and an average random orientation of an “infinite number” of crystallites in a > 1 cm^3 specimen. X-ray diffraction patterns are unique for each crystalline mineral and are used for identification, usually by automated comparison with an electronic library of known and indexed patterns [30]. Although the main mineral phases in cement clinkers and anhydrous cement powders are relatively simple to identify by XRD, the identification of hydrated cement phases in concrete is more challenging. For example, CSH, the main hydration product of Portland cement, due to their poor crystallinity, give weak broad diffraction peaks which tend to be engulfed by the diffraction peaks from CH. Furthermore, a percentage of at least 5% is necessary for the detection of several cement hydrate phases [7].

X-ray diffraction in this study used a Siemens D5000 diffractometer and associated EVA software (V.18.0) and JCPDS PDF-2 (2004) database, to confirm the identity of abundant mineral phases present in the investigated concrete samples (Figure 4(a)). The technique complements the SEM-EDS, EPMA and QEMSCAN® because it is sensitive to crystal structures and can in this way discriminate minerals that are chemically similar or polymorphs (e.g., calcite/aragonite/vaterite). It has also aided in the identification of some forms of concrete deterioration, particularly surface deposits and crack infillings that formed, for example, in response to sulphate attack. Figure 4(b) shows an example diffractogram of a concrete sample using oyster shells as coarse aggregate replacement immersed for 3 months in the coastal waters of the UK. Quartz was the dominant mineral phase as it is the main constituent of the coarse and fine aggregates. Calcite, the main component of the seashells, and CH were identified as minor phases. The noisy background is indicative of a poorly crystalline material, mainly calcium silicate hydrate, the main hydration product of Portland cement.

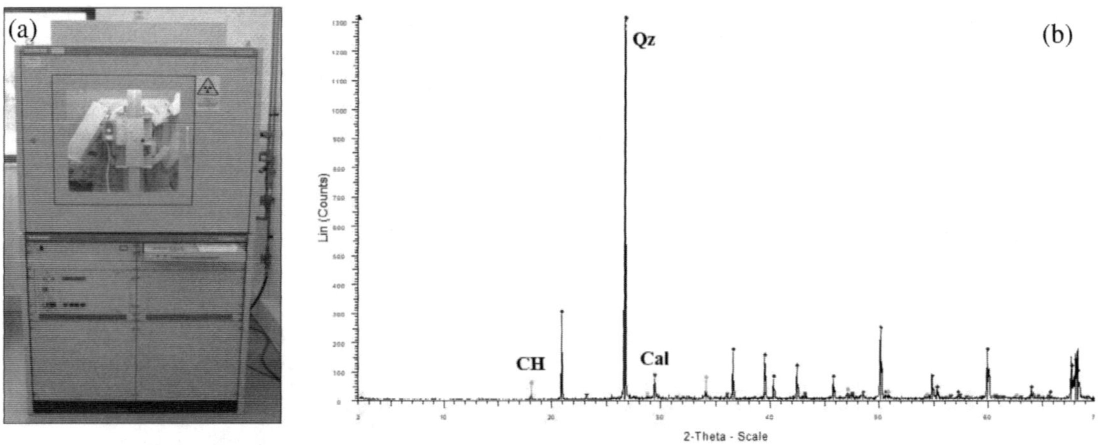

Figure 4. (a) Siemens D5000 diffractometer; (b) XRD diffractogram of a concrete sample made with CEMII, quartz-rich coarse and fine aggregates and oyster shells immersed for 3 months in the coastal waters of the Channel. The amorphous calcium silicate hydrates do not form distinct diffraction signals but indicated from the noisy background. Qz – quartz; CH – calcium hydroxide; cal – calcite.

3. Conclusions

The analytical approach employed for the *MARINEFF* project is a significant advancement that goes beyond the current "state-of-the-art" of established methods. The advantages of the combined analytical approach allow us to better assess the evolution of the concrete materials under marine conditions. The integrated results from optical microscopy, SEM-EDS, EPMA, QEMSCAN® analysis and XRD give unparalleled insight into the variable and complex mineralogy and microtextures of concrete. The study confirms the knowledge of how this material responds when is subjected to an aggressive environment such as the sea, but it also adds further details on the progression of chemical reactions, as well as how these reactions are controlled by the aggregates, fractures and voids. The information range from the quantification of the bulk mineralogy, identification of secondary mineral phases in the cement paste resulting from seawater attack, morphology and distribution of aggregates, seashells and porosity to the presence of cracks and their infilling products, among other. It is clear, that a high degree of detail is critical when it comes to concrete strength and degradation. Such information cannot be achieved from the traditional methods of bulk physical and chemical testing with optical microscopy. We recommend that mineralogical analysis is integrated with data provided through engineering investigations, such as compression strength, porosity, chloride ion ingress tests, and bio-colonization studies.

Acknowledgments

The Authors acknowledge the European cross-border cooperation program INTERREG VA France (Channel) – England co-funded by the European Regional Development Fund for the financial support of the project *MARINEFF* (project 162). QEMSCAN is a registered trademark of FEI Corporation (now part of ThermoFisher Scientific).

References

[1] Falade F 1995 An investigation of periwinkle shells as coarse aggregate in concrete. *Build Environ* **30** 573–77
[2] Cuadrado-Rica H, Sebaibi N, Boutouil, M and Boudart B 2016 Properties of ordinary concretes incorporating crushed queen scallop shells *Mater Struct* **49** 1805–16
[3] Martínez-García C, González-Fonteboa B, Martínez-Abella, F, Carro-López D 2017. Performance of mussel shell as aggregate in plain concrete *Constr Build Mater* **139** 570–83
[4] Mo K H, Alengaram U J, Jumaat M Z, Lee, S C Goh, W I, Yuen C W 2018 Recycling of seashell waste in concrete: *A review Constr Build Mater* **162** 751–64
[5] Foti D and Cavallo D 2018 Mechanical behavior of concretes made with non-conventional organic origin calcareous aggregates *Constr Build Mater* **179** 100–06
[6] Tayeh B A, Hasaniyah M W, Zeyad A M, Yusuf M O 2019 Properties of concrete containing recycled seashells as cement partial replacement: A review *J Clean Prod* **237** 117723
[7] Poole A B, Sims I 2016 Concrete Petrography. A handbook of investigative techniques (CRC Press Taylor & Francis Group, Boca Raton, FL) p 754
[8] Jakobsen U H, De Weerdt K Geiker M R 2016 Elemental zonation in marine concrete *Cement Concrete Res* **85** 12-27
[9] De Weerdt, K, Justnes H, Geiker MR 2014 Changes in the phase assemblage of concrete exposed to sea water *Cem Concr Compos* **47** 53–63
[10] BS 1881-124 2015 Testing concrete – Part 124: Methods for analysis of hardened concrete BSI Standard Publication, The British Standards Institution p 36
[11] The Concrete Society (2010) Concrete petrography: an introductory guide for the non-specialist Technical Report 71, Camberley: *The Concrete Society* p 100
[12] Walker H N, Lane D S and Stutzman P E 2006) Petrographic Methods of Examining Hardened Concrete: A Petrographic Manual. Manual. Revised 2004 U.S. Department of Transportation, Federal Highway Administration Publication N. FHWA-HRT-04-150 USA p 35
[13] APG SR2 2010 A code of practice for the petrographic examination of concrete Applied Petrography Group SR2 Compiled by M A Eden on behalf of the Applied Petrography Group p 20
[14] Jana D 2006 Sample preparation techniques in petrographic examinations of construction materials: A state-of-the-art review Proceedings of the 28th ICMA Conference on Cement Microscopy (Denver, Colorado) 23–70
[15] ASTM C856/C856M 2020, Standard Practice for Petrographic Examination of Hardened Concrete ASTM International (West Conshohocken, PA) p 15
[16] Reed SJB 2005 Electron Microprobe Analysis and Scanning *Electron Microscopy in Geology* 2nd ed. (Cambridge University Press, Cambridge, UK) p 192
[17] Ramos V 2013 Characterization of the potential reactivity to alkalis of Portuguese aggregates for concrete PhD Thesis (Universidade do Porto, Portugal) p 417
[18] Egerton R F 2007 Physical principles of electron microscopy. An introduction to TEM, SEM and EAM (*Springer*) p 202

[19] Papike J J 1987 Chemistry of the rock-forming silicates: ortho, ring, and single-chain structures *Rev Geophys* **25** 1483-526

[20] Papike J J 1988) Chemistry of the rock-forming silicates: multiple chain, sheet, and framework structures *Rev Geophys* **26** 407–44

[21] Deer W A, Howie R A, Zussman J 1992 An introduction to the rock forming minerals 2nd excerpted student edition (Longman, Burnt Mill) p 696

[22] Gottlieb P, Wilkie G, Sutherland D, Ho-Tun E, Suthers S, Perera K, Jenkins B, Spencer S, Butcher A, Rayner J 2000 Using quantitative electron microscopy for process mineralogy applications JOM **52** 24–5

[23] Rollinson G 2021 Automated Mineralogy by SEM-EDS Encyclopedia of Geology 2nd edition (Academic Press) 546−53

[24] Pirrie D, Butcher A R, Power M R, Gottlieb P, Miller G L 2004 Rapid quantitative mineral and phase analysis using automated scanning electron microscopy (QemSCAN); potential applications in forensic geoscience Geological Society, London, *Special Publications* **232** 123–36

[25] Goodall W R, Scalesa P J, Butcherb A R 2005 The use of QEMSCAN and diagnostic leaching in the characterisation of visible gold in complex ores *Miner Eng* **18** 87–86

[26] Pirrie D, Rollinson G K 2011 Unlocking the applications of automated mineral analysis *Geol Today* **27** 226–35

[27] Scott P W, Rollinson G K 2016 Crushed rock aggregates: their mineralogy and textures using automated scanning electron microscopy Proceedings of the 18th Extractive Industry Geology Conference 2014 and technical meeting 2015, EIG Conferences Ltd eds E Hunger and T J Brown 49–68

[28] Pascoe R D, Power M R, Simpson B 2007 QEMSCAN® analysis as a tool for improved understanding of gravity separator performance Min Eng **20** 487–495

[29] Sutherland D 2007 Estimation of mineral grain size using automated mineralogy Min Eng **20** 452−60

[30] Klug H P and Alexander L E 1974 X-Ray Diffraction Procedures for Polycrystalline and Amorphous Material (Wiley Intersciences, New York) p 960

[31] RILEM AAR-1.1 2013 Detection of potential alkali-reactivity, Part 1: petrographic examination method for aggregates Working Group Draft, August 2013

IOP Conf. Series: Materials Science and Engineering 1245 (2022) 012014 doi:10.1088/1757-899X/1245/1/012014

Marine Biofilm effect on mortar surface layer.

M Georges[1], A Bourguiba[1], B Maestracci[2], S Gascoin[2], D Goux[3], D Chateigner[2] and M Boutouil[1]

[1] Normandie Université. Laboratoire ESITC - ESITC Caen. 1 rue Pierre et Marie Curie. 14610 Epron, FR.
[2] Normandie Université. CRISMAT UMR CNRS n°6508. ENSICAEN. IUT Caen. Université de Caen Normandie. 6 boulevard Maréchal Juin. 14050 Caen, FR.
[3] Centre de Microscopie Appliquée à la Biologie, SF 4206 Interaction Cellule-Organisme Environnement (ICORE), UNICAEN ; et CRISMAT, Normandie Univ, ENSICAEN, UNICAEN, CNRS, CRISMAT, 14050 Caen, FR.

Abstract. The objective of this study is to evaluate the effect of biofilm development on the surface layer of a mortar. This work aims at understanding the modifications occurred within the cementitious matrix such as its composition as its microstructure following the bio-colonization. These properties were studied after 360 days of immersion in two immersion media (Artificial Seawater (AS) and Natural Seawater (NS)) using ATG/DTG, XRD analyses and SEM observations. It was found that the biofilm caused the dissolution of the portlandite deeper than when the mortar was not colonized (AS medium). Moreover, a strong carbonation was observed in the first millimeters of the matrix. Two polymorphs of calcium carbonate were particularly identified. Aragonite is the dominant form present on the surface of the mortar immersed in NS. Calcite and magnesian calcite were also observed but after bio-colonization of the mortar. Thus, the development and the biological activity of the biofilm seems to modify the properties of the mortar and could control the precipitation of calcium carbonates on its surface.

1. Introduction

Cementitious materials are bio-receptive in the marine environment [1,2]. Indeed, these substrates are immediately covered by a thin layer of biofilm, which constitutes the first phase of bio-colonization. The biofilm is composed of a succession of benthic communities (mainly diatoms [3]) that promote the second phase of colonization with the attachment of macro-algae and other sessile organisms [3,4].
The objective of this study is to study the interaction between a cementitious material and a marine biofilm. The substrate used is a mortar. When the latter is immersed in the marine environment, it is subjected to different types of physical, chemical [7-10] and biological [11] attacks that can have an impact on its matrix. Indeed, the development of a biofilm on the surface of the mortar can lead to its bio-deterioration. Hueck, (1965) [12] defined biodeterioration as "*any undesirable change in the properties of a material caused by the vital activities of organisms*". Several studies [13-15] have shown that some ions of the cementitious matrix released into the water column can be assimilated by microphytobenthos and macroalgae for their metabolism.
Mesocosm and laboratory studies were conducted to investigate the evolution of the microstructure and composition of the surface layer of the mortar once biocolonized after 360 days of immersion. The following techniques were used jointly for this work: Thermogravimetric analysis (TGA/TDG), Scanning Electron Microscope (SEM) observation and X-ray Diffraction (XRD).

2. Materials and Methods

2.1. Experimental Setup

For this study, mortar samples were made from CEM II/A-LL cement and siliceous alluvial sand (Ø 0/2 mm). The samples studied (Ø110mm/H50mm and Ø110mm/H110mm) were prepared according to the NF EN 196-1 standard and then exposed to a Relative Humidity (RH) of 100% for 24 hours, before being demoulded and stored in water for 90 days at room temperature. After 90 days of curing at 100% RH, the samples were immersed in natural seawater (NS) (mesocosm at the CREC station in Luc-sur-Mer, FR). To distinguish the effect of chemical attack of the seawater from the effect of bio-colonisation on the mortar, samples were also immersed in the laboratory in artificial seawater (AS) obtained from the dissolution of Instant Ocean© salt, Spectrum Brand dosed at 33.3 $g.L^{-1}$ [16].

2.2. Evolution of the microstructure and composition of the mortar surface layer

2.2.1. Thermogravimetric analysis (TGA/TDG). Thermogravimetric analyses provide information on the evolution of the mineralogical composition of the material. These analyses were carried out using an STA 449 F5 (Netzch). Tests were carried out over a temperature range of 20°C to 1000°C, with a heating rate of 10°C/min and a nitrogen atmosphere of 50 mL/min. The tests were carried out on powder samples obtained after sampling the surface layer (0 to 10 mm deep every two millimetres) of the reference mortar (after 90 days of curing) and on samples after their immersion (AS and NS medium). From the thermogravimetric analyses, it is possible to estimate the amount of portlandite and calcium carbonates present in the mortar.
The amount of initial portlandite is determined from the measurement of the amount of water released by mass loss according to the equation (1) :

$$m_{portlandite} = \frac{M_{portlandite} \times m_{water}}{M_{water}} \quad (1)$$

With $m_{portlandite}$ the mass of portlandite (g), $M_{portlandite}$ the molar mass of portlandite (73 $g.mol^{-1}$), m_{water} the mass of water (g) lost during the temperature rise (between 400 and 600°C) specific to the decomposition of the portlandite and M_{water} the molar mass of the water (18 $g.mol^{-1}$).
As with portlandite, it is possible to estimate the amount of calcium carbonate present in the material using the loss of mass from carbon dioxide according to the equation (2) :

$$m_{CaCO_3} = \frac{M_{CaCO_3} \times m_{CO_2}}{M_{CO_2}} \quad (2)$$

With m_{CaCO_3} mass of $CaCO_3$ (g), M_{CaCO_3} the molar mass of $CaCO_3$ (100 $g.mol^{-1}$), m_{CO_2} the mass of carbon dioxide (g) lost during the temperature rise specific to the decomposition of the $CaCO_3$ (between 600°C and 1000°C) and M_{CO_2} the molar mass of CO_2 (44 $g.mol^{-1}$).

2.2.2. Scanning electron microscopy (SEM). The morphology of the main mineral components of the mortar, at the micrometric scale, was analysed using a JEOL 7200 LV Scanning Electron Microscope (SEM), an instrument of the CRISMAT laboratory (Caen, France). The samples (Ø110mm/H110mm) were fixed with 2.5% glutaraldehyde for a few days in 0.2M sodium cacodylate buffer pH 7.4 in the presence of 7% sucrose, then rinsed in this buffer. The samples were then dehydrated by progressive alcohol baths and dried under vacuum (200mbar) after two HMDS (hexamethyldisilazane) baths. The samples are glued on a SEM pad with a carbon glue tape to evacuate the charges.

2.2.3. X-ray diffraction (XRD). The surface of the samples (Ø110mm/H110mm) was characterised using the TFS 4-circle diffractometer (Ecocorail) equipped with a CPS 120 detector and a Cu K wavelength

of Cu $K\alpha_{12}$ [17]. The diffractometer is calibrated on a NIST LaB6 standard powder (srm 660a). The latter was used to calibrate the instrumental contribution [18]. An angle of incidence w=3° and an acquisition time of 12 hours were used. The footprint of the beam size on the sample surface remains within the 1-2 mm^2 during the measurement.
The identification and quantification of the phase was done by referring to the Crystallography Open Database (COD) [19] and by performing a Rietveld refinement operated using MAUD [20].

3. Results and Discussions

3.1. Mortar composition from thermogravimetric analysis (ATG/DTG)

Initially, the mass percentage of portlandite in the surface layer (from 0 to 10 mm) of the mortar studied has an average value of 6.3% (Figure 1a). After immersion, the amount of portlandite decreases only in the first two millimetres in the mortar immersed in the AS medium, whereas it continues to dissolve deeper in the mortar immersed in the NS medium reaching 4.4% between 6 and 8 mm depth. As a result, the microphytobenthos accelerates the dissolution of the portlandite in the cement matrix
As for the mass percentage of calcium carbonates in the mortar, the average value is 2.3%. Once the mortar is immersed in the two media studied, the quantity of calcium carbonate increases only in the first two millimetres of depth (Figure 2b). The increase in $CaCO_3$ may be due to the carbonation of the portlandite following its dissolution (Figure 2a). This carbonation is greater at the surface of the mortar immersed in the AS medium (12.2%) than when it is colonised (6.8%). This difference can be explained on the one hand by the decrease in the quantity of Ca^{2+} ions in the medium following its assimilation by the microphytobenthos. On the other hand, the fixation of dissolved CO2 during photosynthesis reduces its availability in the water column. The form of the neoformed $CaCO_3$ will be more precisely identified after the analysis of the samples by XRD.

IOP Conf. Series: Materials Science and Engineering 1245 (2022) 012014 doi:10.1088/1757-899X/1245/1/012014

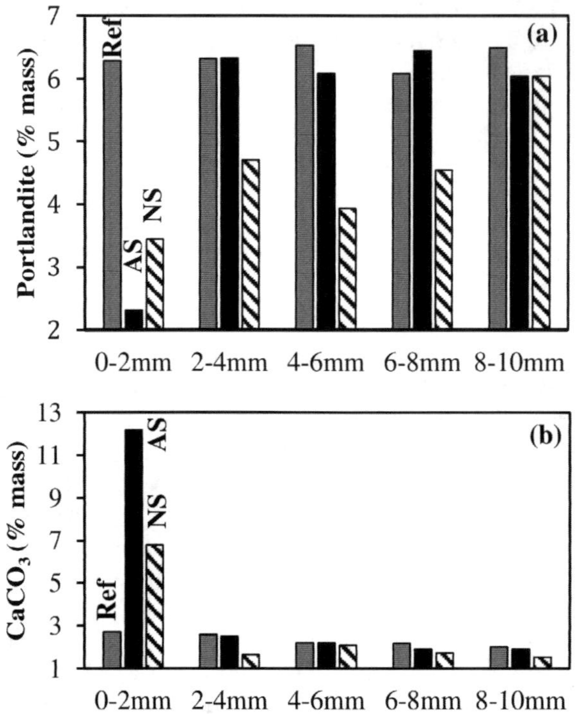

Figure 1. Mass percentage (%) of the amount of portlandite (a) and the amount of calcium carbonates (b) as a function of depth (0 to 10 mm) in the reference mortar (Ref) and the mortar immersed after 360 days in artificial seawater (AS) and in natural seawater (NS).

3.2. Scanning Electron Microscopy (SEM)

Figure 2a show the appearance and texture of the material studied before its immersion. The facies of the different phases initially found in the cement paste are difficult to distinguish. However, we can clearly observe C-S-H gel. Once immersed in the AS medium, the mortar shows orthorhombic crystals on its surface, characteristic of aragonite (Figure 2b) [21]. he precipitation of aragonite is due to the strong presence of magnesium ions in the sea water [22]. Changes in the morphology of the neo-formed carbonates are observed on the surface of the mortar immersed in natural seawater (NS). Indeed, aragonite was found but in the form of cauliflower and another phase which can be similar to calcite (Figure 2c et d). Further XRD analysis will allow the specification of the calcium carbonate polymorph identified.

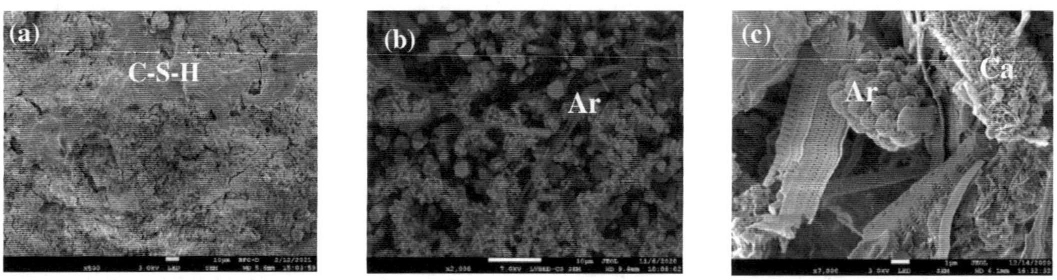

Figure 2. SEM observations of a mortar: reference (before immersion) (a); immersed after 360 days in Artificial Seawater medium (b); in Natural Seawater medium (c) (C-S-H : gel de Calcium-Silicate-Hydrate ; Ar : Aragonite ; Ca : Calcite).

3.3. Scanning Electron Microscopy (SEM)

Mineralogical analysis by X-ray diffraction (XRD) was carried out in order to compare the nature of the crystallised mineral phases present on the surface of the non-immersed mortar (reference) and immersed in the two media studied (AS, NS). The mineralogical composition of the surface of the reference sample consists of quartz and calcite (Tableau 1). Calcite represents 95.7% of the sample. This result means that the reference sample was carbonated before its immersion in the different media studied. This carbonation has a positive effect on the kinetics of bio-colonisation since some studies [23,24] have shown that it improves the initiation of biofilm on the surface of the cementitious material.

In AS, a decrease of calcite in favour of aragonite is observed (confirmed by SEM observations (Figure 2b)). Abiotic factors (high concentration of Mg^{2+} ion) of the environment can explain the preponderance of aragonite.

In the NS environment, three forms of calcium carbonates were observed. The presence of magnesian calcite is explained by the incorporation of Mg^{2+} ions in the crystal structure of calcite [25]. These results show that the development of biofilm on the mortar surface may have caused a change in the crystal lattice structure of the calcite.

Table 1. Percentage (%) of crystallised mineral phases on the surface of a reference mortar and after 360 days of immersion in Artificial Seawater (AS) and Natural Seawater (NS).

	Reference	EA	EN
Halite	-	15.3 (+- 0.5)	-
Aragonite	-	62.8	38.6 (+- 2.7)
Calcite	95.7	21.8 (+- 1.9)	30.5 (+- 2.2)
Calcite Mg	-	-	28.3 (+- 2.1)
Quartz	4.23 (+- 0.1)	-	2.5

4. Conclusions

The different techniques applied in this study provide complementary results necessary to understand the interactions taking place within the mortar surface layer once the material is colonised.

The development and biological activity of the biofilm caused an acceleration of the leaching of portlandite within the cement matrix. Calcium ions released into the medium can then be assimilated by the microphytobenthos.

Aragonite is the majority form on the surface of mortar immersed in artificial seawater. In natural seawater, other forms of calcium carbonates were observed such as calcite and magnesian calcite. The biofilm was able to control the precipitation of calcium carbonate on the mortar surface.

Acknowledgements

This work was carried out with the support of the Interreg FCE MARINEFF Project. Many thanks to the University of Caen for allowing us to immerse our samples in the CREC of Luc-sur-Mer. The authors would also like to thank the funders and all the project partners for their support.

References

[1] Hayek M, Salgues M, Habouzit F, Bayle S, Souche J-C, Weerdt K D and Pioch S 2020 La bioréceptivité de matériaux cimentaires dans l'eau de mer : mécanismes, facteurs agissants et conséquences *Revue Paralia* **13**

[2] Marquez-Peñaranda J F, Sanchez-Silva M, Husserl J and Bastidas-Arteaga E 2016 Effects of biodeterioration on the mechanical properties of concrete *Mater Struct* **49** 4085–99

[3] Salta M M, Wharton J A, Blache Y, Stokes K R and Briand J-F 2013 Marine biofilms on artificial surfaces: structure and dynamics. *Environmental microbiology* **15** 2879–93

[4] Hung O, Thiyagarajan V, Zhang R, Wu R and Qian P-Y 2007 Attachment of Balanus amphitrite larvae to biofilms originated from contrasting environments in Hong Kong *Marine Ecology-progress Series* **333** 229–42

[5] Hanlon N, Firth L B and Knights A M 2018 Time-dependent effects of orientation, heterogeneity and composition determines benthic biological community recruitment patterns on subtidal artificial structures *Ecological Engineering* **122** 219–28

[6] Pioch S, Relini G, Souche J C, Stive M J F, De Monbrison D, Nassif S, Simard F, Allemand D, Saussol P, Spieler R and Kilfoyle K 2018 Enhancing eco-engineering of coastal infrastructure with eco-design: Moving from mitigation to integration *Ecological Engineering* **120** 574–84

[7] Jakobsen U H, De Weerdt K and Geiker M R 2016 Elemental zonation in marine concrete *Cement and Concrete Research* **85** 12–27

[8] Qu F, Li W, Dong W, Tam V W Y and Yu T 2021 Durability deterioration of concrete under marine environment from material to structure: A critical review *Journal of Building Engineering* **35** 102074

[9] Ting M Z Y, Wong K S, Rahman M E and Meheron S J 2021 Deterioration of marine concrete exposed to wetting-drying action *Journal of Cleaner Production* **278** 123383

[10] Yi Y, Zhu D, Guo S, Zhang Z and Shi C 2020 A review on the deterioration and approaches to enhance the durability of concrete in the marine environment *Cement and Concrete Composites* **113** 103695

[11] Márquez J F, Sanchez-Silva M and Husserl J 2013 Review of reinforced concrete biodeterioration mechanisms *Proceedings of the 8th International Conference on Fracture Mechanics of Concrete and Concrete Structures, FraMCoS 2013* 2088–96

[12] Hueck H J 1965 The biodeterioration of materials as a part of hylobiology *Material und Organismen* **1** 5–34

[13] Hughes P, Fairhurst D, Sherrington I, Renevier N, Morton L H G, Robery P C and Cunningham L 2013 Microscopic study into biodeterioration of marine concrete *International Biodeterioration & Biodegradation* **79** 14–9

[14] Javaherdashti R, Nikraz H, Borowitzka M, Moheimani N and Olivia M 2009 On the impact of algae on accelerating the biodeterioration/biocorrosion of reinforced concrete: A mechanistic review *European Journal of Scientific Research* **36** 394–406

[15] Jayakumar S and Saravanane R 2010 Biodeterioration of Coastal Concrete Structures by Marine Green Algae *International Journal of Civil Engineering* **8** 352–61

[16] Atkinson M J and Bingma C 1997 Elemental composition of commercial seasalts *Journal of Aquariculture and Aquatic Sciences, VIII, 2,* 39-43.

[17] Morales M, Chateigner D, Lutterotti L and Ricote J 2002 X-ray Combined QTA Using a CPS Applied to a Ferroelectric Ultrastructure *Materials Science Forum - MATER SCI FORUM* **408–412** 113–8

[18] Caglioti G, Paoletti A and Ricci F P 1958 Choice of collimators for a crystal spectrometer for neutron diffraction *Nuclear Instruments* **3** 223–8

[19] Gražulis S, Daškevič A, Merkys A, Chateigner D, Lutterotti L, Quirós M, Serebryanaya N R, Moeck P, Downs R T and Le Bail A 2012 Crystallography Open Database (COD): an open-access collection of crystal structures and platform for world-wide collaboration *Nucleic Acids Res* **40** D420–7

[20] Lutterotti L, Matthies S, Wenk H-R, Schultz A S and Richardson J W 1997 Combined texture and structure analysis of deformed limestone from time-of-flight neutron diffraction spectra *Journal of Applied Physics* **81** 594–600
[21] Palin D, Wiktor V and Jonkers H M 2015 Autogenous healing of marine exposed concrete: Characterization and quantification through visual crack closure *Cement and Concrete Research* **73** 17–24
[22] Berner R A 1975 The role of magnesium in the crystal growth of calcite and aragonite from sea water *Geochimica et Cosmochimica Acta* **39** 489–504
[23] Hayek M, Salgues M, Habouzit F, Bayle S, Souche J-C, Weerdt K D and Pioch S 2020 L'influence de la carbonatation sur la biocolonisation de matériaux cimentaires dans le milieu marin *Matériaux & Techniques* **108** 202
[24] Natanzi A S, Thompson B J, Brooks P R, Crowe T P and McNally C 2021 Influence of concrete properties on the initial biological colonisation of marine artificial structures *Ecological Engineering* **159** 106104
[25] Loste E, Wilson R M, Seshadri R and Meldrum F C 2003 The role of magnesium in stabilising amorphous calcium carbonate and controlling calcite morphologies *Journal of Crystal Growth* **254** 206–18

Preliminary evidence of early Reef Cube® enhancements to mobile benthic ecosystems

S J S Hickling and J A J Murphy

ARC Marine Ltd., Brixham Laboratory, Freshwater Quarry, Brixham, Devon TQ5 8BA, UK

Corresponding author emails: sam.hickling@arcmarine.co.uk and james.murphy@arcmarine.co.uk

Abstract. Reef Cubes® are ecological engineering units, intended to restore biodiversity while serving functions to marine engineers. Ecological engineering must be backed by quantitative scientific evidence. Here, preliminary results of the Reef Enhancement for Fisheries at Aquaculture Sites (REFAS) project are presented. 11 Clusters of Reef Cubes® were deployed in June 2021. Sampling was undertaken before and after in the Spring and Autumn of 2021, in daylight and at night-time. Guided Baited Remote Underwater Video Station (BRUVS) surveys were completed with a modified BlueROV2. All mobile benthic and demersal species were identified during a 30-minute window and quantified in terms of Max-N. Biodiversity indices, rarefaction curves and multivariate community compositions were compared at reef and non-reef sites before and after build. Average species richness, Pielou's evenness, Simpson's and Shannon's indices were similar between reef and non-reef sites. Seasonal increases of total abundance and total species were found, and these were significantly greater at reef sites. There were also significant differences in the community compositions of the reef and non-reef sites that were not observed before the build. These results are significant for the implications of using Reef Cubes® and indicate that Reef Cubes® could be a potential tool for biodiversity remediation.

1. Introduction

The world is undergoing a biodiversity crisis. Species extinctions are increasing rapidly and predictions suggest that the world is on the verge of a new mass extinction event, at the hands of modern anthropogenic impacts [1]. Marine spaces are not exempt from the impacts of industrial man kind and have undergone significant changes in recent history; with an overarching decline in biodiversity [2]. One such change is the mass alteration of sea floor habitats, following a general trend of degradation from structurally complex to simple states [3]. The characteristics of habitats is a key aspect of the surrounding ecosystems and thus the degradation of marine habitats is closely intertwined with biodiversity losses. Indeed, industrial extraction of fishery resources has been a key cause of habitat degradation and simultaneously depletes populations of marine fauna, with habitat damaging methods often being unselective in their extraction. Construction in the marine environment has, historically, paved over large swathes of complex biogenic habitats, especially in the intertidal zone, replacing natural habitat systems with simple artificial grey and rock infrastructure [4]. There is, however, the possibility that construction can be used as a tool for biodiversity restoration and for protecting existing vulnerable habitats [5–7]. Indeed, the field of ecological engineering is continually developing design recommendations to improve the effects of construction with opportunities for significant positive effects to biodiversity as by-products of ongoing projects [5,8].

The restoration of structural complexity in the marine environment is a key step required towards the renewal of marine biodiversity. Technically restoration projects should use accurate pre-impact baselines with which to reference the success of projects [9]. Pre-impact baselines for benthic habitats however are rarely available as pervasive anthropogenic impacts often pre-date the majority of survey efforts [10,11]. There is therefore an imperative to rebuild benthic habitat complexity without a strict historic baseline to target, with a diversification in project aims from geographically accurate restoration to the wider restoration of habitat complexity and biodiversity at large [12]. Geographically accurate restoration projects should still be pursued as far as is possible but there could also be an acceptance that the positive effects of engineering projects can be a potential remediation for widespread historic benthic habitat degradation, with the ultimate effects on biodiversity balancing out historic trends. Strictly the terminology for this effect and process is yet to be decided but there are a variety of candidates: restoration, regeneration, renewal, rewilding, net gain and enhancement etc. It is this positive effect that is, to date, largely unrecognized in planning processes, especially in the United Kingdom. There is no doubt that the urban sprawl of mankind into marine spaces represents a serious harm to biodiversity but it also represents an opportunity. Artificial habitats in the marine environment are here to stay and will increase, they must be optimized with biodiversity in mind to not only reduce their ongoing impacts but to make the most of this powerful tool for ecosystem engineering.

Acceptance of the potential that man-made structures have, means that a wide variety of ongoing projects with set motivation for funding to humanity can be used as a vehicle for biodiversity restoration. This work around means that the existing ongoing impacts of marine construction for projects such as intertidal coastal defence works, aquaculture projects, harbour construction and offshore energy generation can be exploited and used as a vehicle for positive change. The narrative could be rewritten so that where projects are only perceived now for their potential negative impacts, they could also be seen in the future as a means for regenerative opportunity – when designed correctly. It's therefore imperative that ecological engineering is integrated with engineering industries in the marine space for two major positive effects: restoring habitat complexity on a wide geographic basis and improving the ongoing impacts of engineering on the seafloor by increasing the quality of the habitat deployed. Such integration must come with a warning, however, that ecological engineering should not be used to green wash development plans and accelerate the spread of anthropogenic infrastructure unnecessarily [13].

ARC Marine Ltd. is an eco-engineering company based in the Southwest UK. Their primary product is a patented artificial reef unit called the Reef Cube®, which is a basic building block in a variety of sizes. Reef Cubes® can be used alone or in combinations, to suit different specific engineering or ecological applications. The key feature of the Reef Cube® is that it contains a spherical chamber with passages in each side. The application of voids such as these has been found to dramatically improve the diversity and productivity of ensuing ecosystems around structures and thus gives Reef Cubes® an ecological advantage over standard competitors [14–16].

One such application for Reef Cubes® is in conjunction with aquaculture sites either to support infrastructure as eco-mooring bases or as an intentional habitat enhancement in a holistic multitrophic aquaculture design. The benthic environment is yet a largely unexplored element of aquaculture facilities, especially in addition to line grown shellfish farms. There is, however, evidence of benefits to organisms of commercial value for human consumption such as crab and lobsters below line grown mussel farms; and in the case of lobsters this is likely due to the increase of shelter provision [17]. The combination of artificial reefs with aquaculture sites could be a means of promoting multitrophic aquaculture development by intentionally altering the benthic habitats. It could also be an excellent opportunity to restore biodiversity in the marine space, where impacts are already planned to occur. Indeed, there are calls in the scientific community for greater research of the use of artificial reefs with multi-use aquaculture sites for productivity and conservation purpose [18]. At the least the incorporation of Reef Cubes® could improve the existing impacts that aquaculture facilities have by improving the design of gravity mooring bases, so that a more complex habitat is introduced to the environment.

The Reef Enhancement of Fisheries at Aquaculture Sites (REFAS) project aims to elucidate the effects of Reef Cubes® in combination with aquaculture by researching the use of Reef Cubes® nearby to an innovative scallop farm in Torbay, Devon, UK. Funded by the Seafood Innovation Fund (SIF) through the UK Centre for Environment Fisheries and Aquaculture (CEFAS) the project has undergone two phases. During the first phase a feasibility reef was deployed to check the construction and deployment methods, followed by a long-term reef research site deployed nearby in the second phase. Large Reef Cube® clusters were deployed that could be used for the mooring of research vessels, to emulate the ecological effects of using such designs for moorings at aquaculture facilities. A key aim of the reef construction was to enhance the biodiversity of the area, to ascertain if Reef Cubes® can indeed be used as a biodiversity restoration tool alongside aquaculture construction in the marine environment. Here, a preliminary assessment of this aim and the changes to the mobile demersal and benthic community after 4 months of deployment are described.

2. Methods

2.1. Test Site

Two sets of Reef Cubes® were deployed for project REFAS. A pilot reef was built in 2020 called REFAS 1 and a long-term test reef was built 100m away in 2021 called REFAS 2. REFAS 2 is the subject of this study and the primary site of on-going research. Both sites are located in the North of Torbay, Devon, UK; to the North of a lantern-style scallop aquaculture site owned and operated by the Scallop Ranch Ltd.

The REFAS 2 site is in a box with the centre at 50°26.967'N, 3°29.595'W. The REFAS 1 site is approximately 100m to the South-West. The boundary of the Scallop Ranch is approximately 200m to the South and the site also lies to the South of the Torbay Marine Conservation Zone. Chart datum is 11m, maximum tidal height in the area is 5.4m. The sea floor is a featureless plain of sandy mud as revealed by pre-build bathymetry surveys. The area approximately 200-300m around the site is uniform in depth. The latest habitat classification available at the site is circalittoral sandy mud; with circalittoral and infralittoral sandy mud habitats persisting for most of the open Torbay area and surrounding region [19]. Two pinnacles of rock exist over 500m away: East Shag Rock to the Northwest and Thatcher Rock to the Northeast. Each has some spatially limited subsea rock habitat surrounding the pinnacle. To the North, the coast is lined with shingle habitat.

2.2. Reef Design

11 clusters of Reef Cubes® were deployed on the 2nd June 2021 by Jenkins Marine Ltd, using the deployment vessel Stour. There was a deviation from the planned cluster locations on average of 20m in a North Easterly direction. The clusters consisted of 6 types. 5 of the types each have 2 replicates whilst there is only 1 replicate of the final type (See appendix table 1 for full list of Reef Cube® clusters). The cluster types were specified to include a variety of Reef Cube® sizes to increase the complexity of deployed habitats and some clusters contained Reef Cubes® inside larger Reef Cubes® for the same purpose. The reef site design was specific to ensure equidistance of at least 20m between each cluster. Each cluster had a marker buoy attached by a set of chain brothers, which were also used to deploy the clusters.

2.3. Survey Methods

The REFAS 2 site was surveyed before and after build, in the Spring (9th-18th March, 2021) and Autumn seasons, (18th September – 9th October, 2021), during daylight hours (8:30 to 17:30) and nighttime hours (18:30 to 4:30). Surveys were performed via visual censuses using a modified BlueROV2 as a targeted baited camera station.

The study design followed a Before After Gradient (BAG) style [20]. There is evidence to suggest that this method improves the power of impact assessment studies compared to a Before After Control Impact (BACI) style design [21].

A total of 104 samples were collected. For each sample period, at both day and night, 11 samples were taken at reef sites - one at each cluster location - and 15 samples taken at non-reef sites away from the reef cluster locations. Three samples were taken at increasing distances – 15m, 40m and 130m away from the reef sites along five transects. Samples were alternated between reef and non-reef sites until

all reef sites were completed to ensure that environmental and time variables were equally distributed between reef and non-reef sites.
The planned location of the 11 reef clusters were used for the reef sites before build and the actual location of the reef clusters was used for the reef sites after build. The area around the reef was divided into 5 equal size sectors to select the locations for non-reef site samples. In each of these sectors a random number generator was used to select a bearing for the direction of a transect line. Transect lines were plotted from the further reef site in that direction with survey stations designated at 15m, 40m and 130m away. If two transect lines from separate sectors were randomly selected within 20 degrees of each other another random number was generated to ensure that transects were separated. NE, SE, S, SW and NW sectors were selected to provide an overview of the area around the reef whilst taking into account impacts from the local aquaculture site to the South and the MPA, local coastline and rocky features to the North. The depth remains consistent in this area around the reef site so that direction is not confounded with depth for each transect. Additionally, the habitat type remains largely the same within the 130m radius from the test site, apart from the REFAS 1 site to the Southwest and scallop farm to the South which were not encountered by any of the non-reef transects.
Visual censuses were performed using a stationary baited setup: a baited remote underwater video station (BRUVS). This allowed for the completion of surveys during tidal states when the ROV would have been unstable and uncontrollable as the system can be weighted or piloted with downward thrust to stay on the seafloor. The camera was oriented horizontally with a slight 10-degree downward view to observe the bait. Two laser points at 10cm apart were used for length measurement. The survey time of the camera system was 30 minutes from stable positioning on the seafloor. The minimum distance between stations was 15m between the reef sites.
The bait species was salmon *Salmo salmar*. 200g of chopped salmon were used for each station. The bait was held in a plastic cage with a mesh size of 1.5cm. The cage was held in front of the ROV by an arm 60cm long.
Video footage was analysed with all mobile fauna identified to the highest taxonomic level possible. Abundance was quantified as Max N – the maximum number of each taxon in one frame within the 30-minute survey.

2.4. Statistical Analyses
Univariate and multivariate analyses were performed. Univariate analyses aimed to assess if there were differences in biodiversity indicators at the site type – reef versus non-reef - after the build and if these differences were present in the Spring, before the reef build. The biodiversity indicators tested were: Species Richness, Pielou's Evenness, total abundance, Shannon's diversity index and Simpson's diversity index. For univariate analyses day and night surveys were analysed together. Total abundance was calculated by summing the Max N value for all species in each sample. Biodiversity indicators were tested using Generalized Linear Models (GLMs) against the period (Spring or Autumn) and sample site type (Reef vs Non-Reef) and with interaction between the two effects. A secondary analysis was also completed with distance from the reef sites (0m, 15m, 40m or 130m) used instead of site type. Rarefaction curves were produced for the four separate combinations of site type and period factors. The curves were produced using PRIMER by calculating the mean of species accumulation curves replotted in random order over 999 permutations – using the 'UGE method' [22]. Rarefaction curves were visually compared using the minimum sample number (22) as a reference point for comparison. GLMs were then run on the curve data points to test differences in the effects of the sample number on accumulated species, by site type and period, with an interaction between the two to see if the reef deployment impacted the total species utilising the reef sites.
Multivariate analyses aimed to assess if the reef deployment had affected the community composition at the reef sites compared to the non-reef sites and if so to elucidate what these differences were. Unbalanced crossed PERMANOVA tests (based on Bray-Curtis similarity matrices and 999 unrestricted permutations of raw data) were used to test for differences in multivariate community compositions. The fixed factors tested were site type (reef or non-reef), distance from reef, test period (Spring or Autumn) and test time (day or night). Crossed effects were included for site type with period;

distance from reef with test period; and site type with period and day/night timing. Following a significant result, pairwise tests were completed to identify significant differences between individual levels of the site type and period factors, analysing the day and night samples separately. Prior to analysis Max N of each taxon was square root transformed to account for scale differences and dominance of certain taxa. Type III Sums of Squares partitioning was used. PERMDISP routine was used to check the homogeneity of multivariate dispersions within factors. For pairwise tests with significant results, SIMPER analysis was completed to elucidate which taxa accounted for the greatest dissimilarity between the reef and non-reef sites in the Spring and Autumn periods, before and after reef build.

Univariate analyses were completed in R [23], figures were produced using the package ggplot2 [24] data was manipulated using the tidyverse set of packages [25]. Production of diversity indices and rarefaction curves, and multivariate analyses were carried out in PRIMER version 6 with the PERMANOVA add-on [26,27].

Figure 1. Photographs of the REFAS Reef Cubes®. a) REFAS 2 Reef Cubes® on transit for deployment 2nd June 2021, b) a Reef Cube® formation at the REFAS 1 site, image taken 7th June 2021, deployed 1st April 2020, c) REFAS 2 Reef Cube® cluster 'Reef 6' 5 days after deployment – image taken 7th June 2021, d) REFAS 2 Reef Cube® cluster 'Reef 6' 100 days after deployment – image taken 15th September 2021, e) Internal chamber of REFAS 2 Reef Cube® cluster 'Reef 6' approximately 100 days after deployment – image taken 15th September 2021.

3. Results

3.1. Univariate Biodiversity Metrics

54 separate taxa were counted, with 32 identified to species level. A seasonal effect on species richness and Shannon's diversity indices was observed, with higher averages in the Autumn than the Spring (p values < 0.005). There was no significant effect of distance from the reef on any of the univariate biodiversity metrics: species richness, Pielou's evenness, total abundance, Shannon's and Simpson's

indices (All p-values >0.05). Similarly, there was no significant effect on species richness, Pielou's evenness, Shannon's and Simpson's indices for either period when comparing the site type – reef versus non-reef (All p-values >0.05). There was, however, a significant effect of the period (t = -2.447, p = 0.01615) and the site type (t = 3.066, p = 0.00279) and an interaction between the two (t = -2.386, p = 0.01893) for the total abundance. The coefficients indicated that the Spring period had a lower overall abundance and that there was a significantly greater increase in the abundance at the reef sites from the Spring to Autumn than the non-reef sites. Mean total abundance at the reef sites was 5.6 (+- 0.7 S.E) in the Spring and 22 (+- 4.4 S.E) in the Autumn. At the non-reef sites mean total abundance was 6.5 (+- 1.2 S.E) in the Spring and 13 (+- 1.2 S.E) in the Autumn.
Rarefaction curves indicated a slightly larger overall mobile species pool at the prospective reef sites than the non-reef sites before build in the Spring and after the reef build in the Autumn (see figure 1). At the smallest sample size (22) the predicted species count was highest at the reef sites in Autumn (30), followed by the non-reef sites in Autumn (24), the prospective reef sites in Spring (19) and the non-reef sites in Spring (17). There was a significant seasonal difference in the rarefaction curves between Spring and Autumn, where Spring samples accumulated a lower species count (t = -2.286, p = 0.025). There was also a significant difference between the reef and non-reef sites, where reef sites accumulated a higher species count (t = 7.401, p = 5.15 e-11), and this effect was significantly lesser in the Spring period before the reef was built (t = -2.893, p = 0.00472).

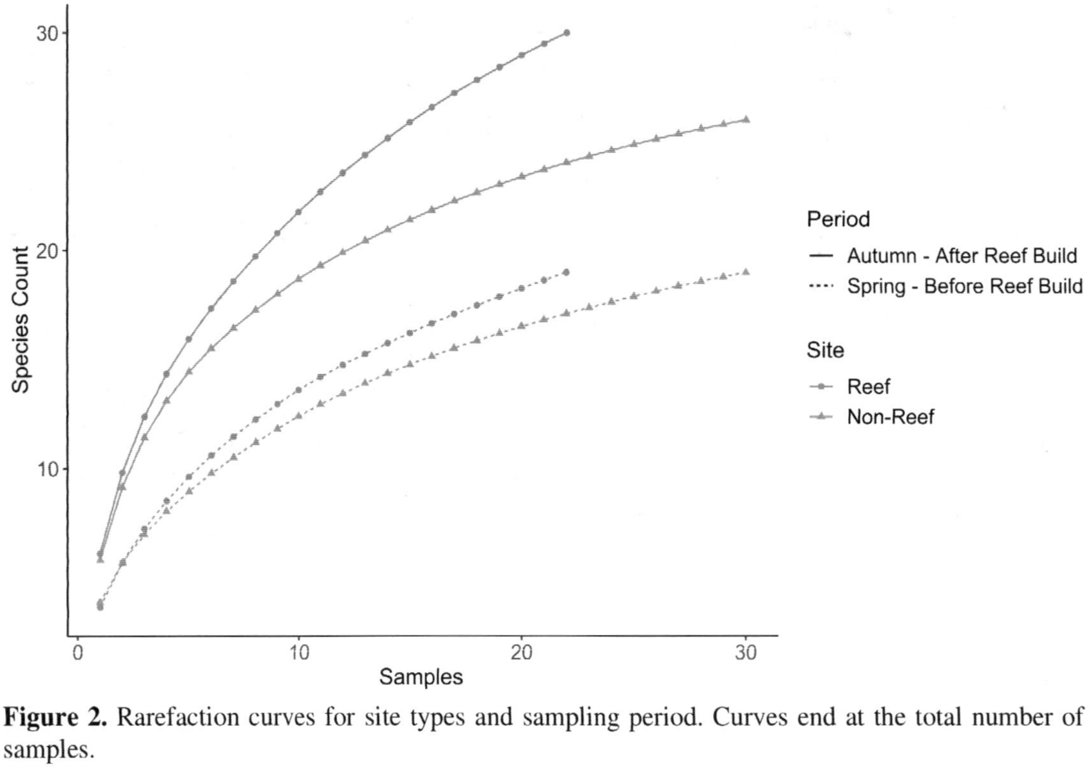

Figure 2. Rarefaction curves for site types and sampling period. Curves end at the total number of samples.

3.2. Multivariate Analyses

Analysis of community similarity revealed that there were significant interactions between the effect of: Reef and Non-reef sites and the sampling period (Pseudo-F 26.226, P-Value 0.001), the distance from the reef and the sampling period (Pseudo-F 10.812, P-Value 0.001) and the differences in day and night communities, reef and non-reef sites and sampling periods (Pseudo-F 1.8887, P (perm) 0.049) (See Appendix Table 2 for full results and Figure 3). Pairwise comparisons revealed that there were no significant dissimilarities during the Springtime - before the reef build - between the prospective reef sites' and non-reef sites' daytime communities (t = 0.81312, P (perm) = 0.636) or nighttime communities (t = 1.067, P (perm) = 0.335). There were, however, significant dissimilarities during the Autumn period - after the reef build - between the reef sites' and non-reef sites' daytime communities (t = 3.3214, P (perm) = 0.001) and nighttime communities (t = 1.7361, P (perm) = 0.01). Pairwise tests also revealed significant seasonal effects of the sampling period on both reef and non-reef sites' day and nighttime communities (All P values <0.005) and there was a greater dissimilarity of the Reef site communities between the Spring and Autumn periods than the non-reef sites.

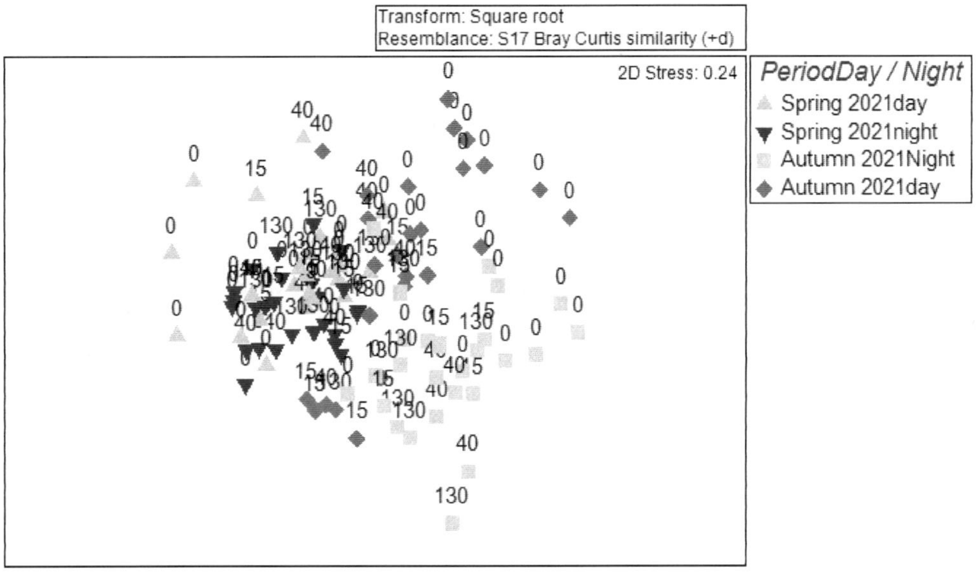

Figure 3. Multi Dimensional Scaling (MDS) Plot. Each point represents one sample. Colours and shapes correspond to combinations of the period and day or night factors. Plot produced from bray curtis similarity plots with an additional dummy species variable. Numbering indicates the distance of the sample site from the reef sites.

SIMPER analyses revealed which species contributed most to dissimilarity between periods and reef or non-reef sites (see figure 4 for summary). There were several species that exhibited considerable differences in abundance between the reef sites and non-reef sites in the Autumn and between the Autumn and Spring periods – pre and post-build – for the reef and non-reef sites respectively. In the Autumn, during the day, at the reef sites, there were notably higher abundances of Poor Cod *Trisopterus minutus*, Pouting *Trisopterus luscus* and Sand Goby *Pomatoschistus minutus*; slightly higher abundances of sea bass *Dicentrarchus labrax*, herring fish Clupidae spp., an unidentified Perchiformes species and the cuttlefish *Sepia apama* which were present on rare occurrences at the reef sites in Autumn but were absent at the non-reef sites and both site types before build. There were also higher abundances of the harbour crab genus (Liocarcinus spp.) and velvet swimming crabs *Necora puber*. In contrast the non-reef sites in the Autumn had higher abundances of brittle stars *Ophiothrix fragilis*, dog whelks *Nucella lapillus* and Hermit crabs (*Paguroidea* spp.).

At night the reef sites in Autumn once again had higher abundances of the gadoids *T. minutus* and *T. luscus* though this difference was less pronounced than in the day. There was also a higher abundance of squid (Teuthida spp.) and common star fish *Asterias rubens* which were absent from the non-reef sites and both site types in the Spring. There was also a very slightly higher abundance of edible crab *Cancer pagurus* though this was limited and there was an overlap in standard errors. In contrast in the non-reef sites at night in the Autumn there was a slightly higher average abundance of horse mackerel *Trachurus trachurus* though with overlap in the standard errors to the reef sites. Both the common

whelk *Buccinum undatum* and dog whelk *Nucilla lapillus* were more abundant in the nighttime at the non-reef sites. The masked crab *Corystes cassivelaunus* was slightly more abundant at the non-reef sites, whilst in the spring this difference was not observed. There was also a higher abundance of hermit crabs again though this was less significant than the day and there was overlap of the standard errors with the reef sites. This difference was also observed in the reef and non-reef sites before build in the Spring. Horse mackerel *T. trachurus* were more abundant at both sites in the Autumn.
There was a slightly higher average abundance of the gurnard *Eutrigla gurnardus*, angular crab *Goneplax rhomboides* and long-legged spider crabs (Macropodia spp.) at the non-reef sites in the day at both Spring and Autumn periods, whilst these species were absent at the reef sites over both periods Spring and Autumn - before and after build.
The most abundant species at any one survey was the poor cod *T. minutus* whereby a large shoal was observed at the Reef 6 site, with 81 counted in one frame. This species was not observed at any non-reef site during the day in the Autumn period and any site at any time in the Spring period; and contributed a high proportion of dissimilarity in SIMPER analyses comparing the reef communities after build with others (>8%). The highest abundance of any one species at a non-reef site was in the Autumn when 15 horse mackerel *T. trachurus* were observed in one frame at night, which is slightly higher than the maximum at a reef site in Autumn which was 9. It's possible that this difference is due to chance upon inspection of the means and standard errors which overlapped.

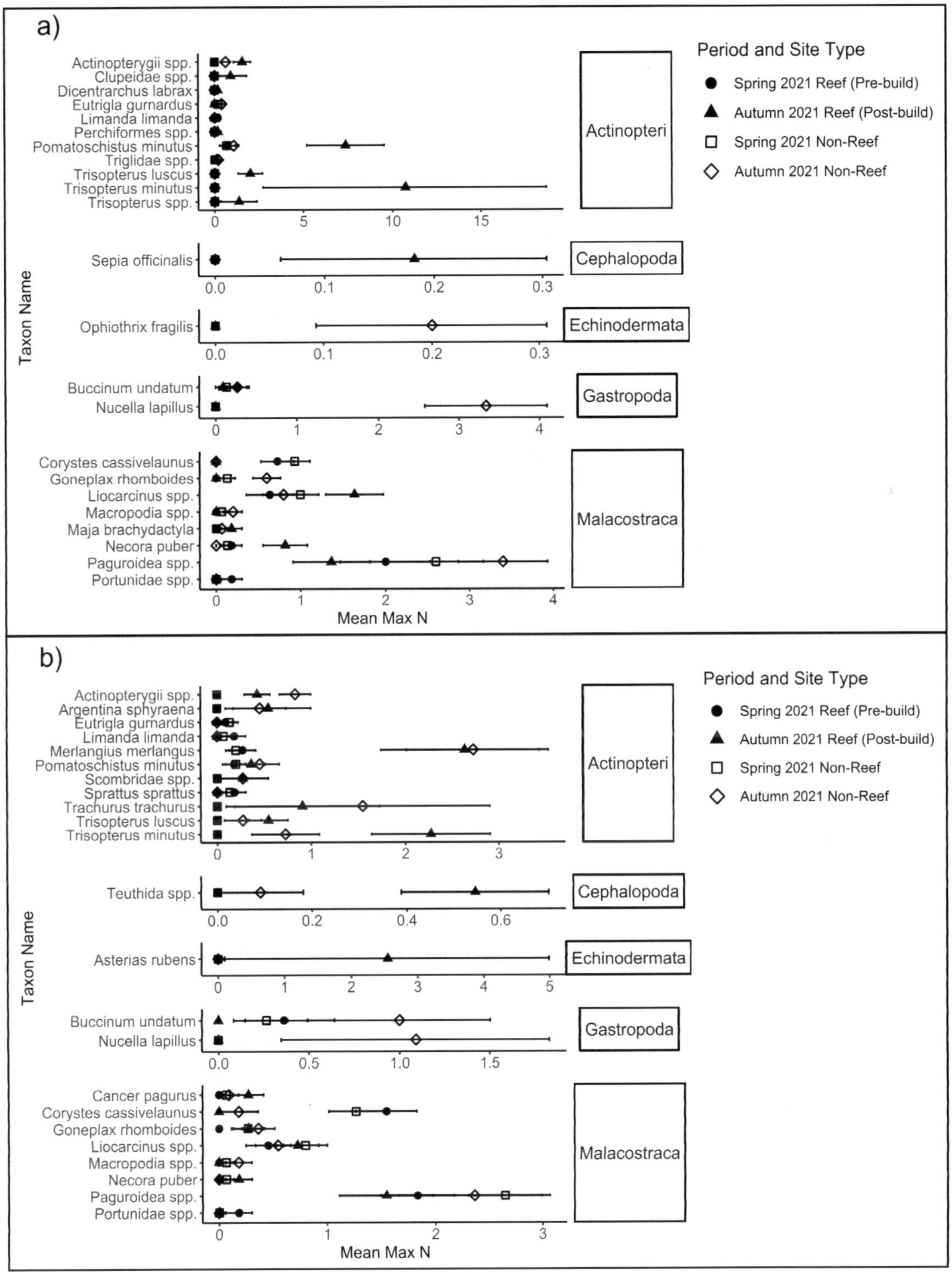

Figure 4. Composite dot plots of average animal abundance in a) day and b) night surveys. The taxonomic classifications are the most specific possible for sightings during surveys. Taxa were selected that contributed most to 90% of the accumulative dissimilarity in SIMPER analyses between reef site types and periods for day and night communities. Charts are organised in to four taxonomic

classes: Actinopteri, Cephalopoda, Gastropoda and Malacostraca; and one taxonomic phylum: Echinodermata. Symbols indicate the combined site type and period factors. Points represent mean Max N (maximum individuals in a single frame) as a conservative estimate of abundance for each species, error bars indicate standard error.

4. Discussion

The analysis revealed potential alterations of mobile benthic ecosystems by the reef deployment through increases in total abundance, the total species pool and the abundances of several particular species, resulting in significant changes to the community composition at the reef sites. Guidance suggests that 'before after' studies should take samples in the same season to avoid seasonal effects in the region confounding the results. Analyses summarized here are therefore preliminary and will be compared with further sampling in the Spring of 2022.

Seasonal effects were observed at all sites and these effects differed between those in the immediate vicinity of Reef Cubes® and those 15m and more away. There were also differences between the day and the night-time communities in the Autumn that were not observed in the Spring and these differences varied between the reef and non-reef sites. There were significant seasonal effects on the overall abundance of all species in the benthic mobile communities and the predicted overall species, with abundance and species pools being higher in the Autumn than the Spring. This difference was of a higher magnitude at the reef sites, which had the greatest average abundance and total species found. As the Reef Cube® structures were deployed between these periods it is plausible that the changes are due to the habitat alteration with the introduction of the Reef Cubes®. The changes appear to be limited to within the direct vicinity of the Reef Cubes® and boundary effects were not clearly identified in this study. These initial changes fit with previous evidence of the attraction effect of artificial reefs, leading to higher abundances and diversity of mobile species [5,28].

Differences in the community compositions at the new reef habitat communities and the surrounding soft sediment habitats were observed after 4 months of deployment, indicating that habitat alteration of the Reef Cube® clusters has produced changes to the ecosystems. The community composition differences were potentially driven by the association of these species with hard or soft bottom habitats. There was one distinct community type present in the area in the Spring before the build but that there were at least four distinct communities in the area with different soft substrate and reef type communities in the day and night. Where there were only soft sediment habitat communities before now there are artificial reef communities and soft sediment communities in the area. Demersal species are highly mobile and thus utilisation and residency of the site will have differed by species; as exemplified by the differences in day and night communities. For example, in the cephalopods cuttlefish were noticeably present at the reef sites in the day, whilst squid were present at night. This points to the different behavioural patterns of the species observed and to the nature of baited video surveying in that species are only likely to be found if actively moving through the environment and more so if attracted by the bait. It is this differential usage of the reef habitat that can be used as a 'rule of thumb' to predict the magnitude of benefits that a species will garner from the habitat, with more closely associated species generally predicted to benefit more so than transient species [29].

Evidence of the reef community alteration can be found by examining the biology of specific species that showed differential abundances at the reef and non-reef sites after the reef was built. Organisms like the common whelk, dog whelk and the hermit crabs were on average more abundant at the non-reef sites. It's possible that the reef deployment displaced these species to varying degrees in the immediate vicinity of the clusters. The results could indicate that these species have a greater affinity to soft sediment habitats or that interspecies interactions are affecting their abundance. The hermit crabs have an association with whelks in the area as they utilise their shells, switching between species as they grow depending on species size variation (personal observation). There doesn't appear to be impacts to these species from 15m away from the reefs and beyond, however, closer examination of boundary effects is required.

The reef sites on the other hand had a higher abundance of species that might be attracted to the reef habitat. For example, many gadoids such as *T. minutus* and *T. luscus* display sheltering behaviour around hard structures and shoals of these two species were observed around the reef structures more so than at the nearby non-reef sites. All the *T. luscus* individuals observed here were juveniles - below mature size - and over 60% of *T. minutus* were also juveniles. Juvenile gadoids such as these are likely to benefit from the introduction of complex hard substrates because of the shelter and feeding benefits [30], indeed *T. minutus* and *T. luscus* are both closely associated to benthic ecosystems and feed as scavengers on the benthos. The shoaling pattern of Trisopterus species has also been observed in the Dutch North Sea around other Reef Cube® structures [31]. Likewise, *Asterias rubens* is a scavenging organism and the higher abundance witnessed here could be stimulated by attraction to greater feeding opportunities around the reef. Sea bass *D. labrax* were sighted at the reef in Autumn but nowhere else. Sea bass are ambush predators with a strong association to hard substrate habitats, it is plausible that sightings of sea bass could have been driven by the presence of these gadoids and the reef structures. Likewise, cuttlefish and squid may have been attracted by feeding opportunities, not only from these small fish but additionally from the higher abundances of Liocarcinus crabs and velvet crabs *Necora puber*. Cuttlefish and squid have been found to attach their eggs to natural fauna and flora growing on hard substrates and artificial substrata such as pots and rope material [32–34]. Eggs have previously been found wrapped around textile ropes attached to Reef Cubes® in Torbay (unpublished results). It could be that these species are attracted to the reef for the potential egg laying opportunities that any attached materials provide, in addition to the potential feeding benefits that might be found. The addition of natural fibre ropes and metal cages to Reef Cubes® was a key design feature in the Dutch North Sea project, with the intention of promoting egg attachment behaviour by cephalopods [35]. Use of the Reef Cubes® as shelter is likely to be more profound for the cuttlefish that are more closely associated to the benthos than the more neritic squid.

The nature of the species differences therefore points to a distinct reef community beginning to be established around the Reef Cube® structures, within the first 4 months of deployment. The effects appeared to be in the immediate vicinity of the Reef Cube® clusters though there is an indication that community composition changes may be on a gradient basis. Future analyses should isolate the sites away from the reef and check for boundary effects from 15m and further away. A greater analysis of juvenile fish across all species could be performed to quantitatively assess the nursery effect of the reef structures. Future research should investigate the effect of the habitat diversification with Reef Cubes® on regional species pools - taking account of both soft and hard substrate habitats. The effect of Reef Cubes® compared to industry standard solutions without eco-engineering designs should also be investigated.

Reef Cubes® as described here can be utilised for applications such as moorings, coastal defence, subsea asset protection or for the changes that they impart on marine ecosystems. This study has provided the first quantitative description of mobile communities utilising Reef Cubes®, 4 months after placement on a muddy sand habitat in the waters of South-West UK. Already a differentiation between the Reef Cube® communities and those in the surrounding sand habitats has been observed. There are two effects on univariate biodiversity metrics – abundance and total species accumulation that can be seen as positive indications of having these new habitats in place. Whilst these effects are positive there is still value in soft sediment habitats and their ecosystems, especially where designated as a feature in Marine Protected Areas. The alteration of habitats must take in to account the factors associated with habitats undergoing alteration, including their uniqueness, geographical extent and biodiversity value. With the general degradation of habitat complexity, habitat quality and biodiversity throughout the world's oceans however it is important to identify tools to promote high biodiversity ecosystems, especially if they can be incorporated into ongoing anthropogenic alterations. These early indications show that Reef Cubes® are a promising tool for the reintroduction of complex habitats into soft substrate areas for the purpose of biodiversity renewal and could be integrated into ongoing construction projects for this purpose. The analysis here provides a basic overview and inevitably the effects of building new habitats is far more nuanced than simply changes in biodiversity indices or changes in animals' abundances; with new structures providing different services to marine organisms in a myriad of ways. Here we have begun to explore some of the basic patterns in biodiversity that occur with Reef Cubes®, without emphasis on behavioural change or the emergence of new functional diversity. It is important that that community changes when utilising Reef Cubes® are understood so that users can optimise them sufficiently to influence ecosystems as desired. An important first step in developing this understanding has therefore been completed, with key species noted in the early stages of reef development.

Acknowledgments

Thanks go to the Seafood Innovation Fund team for funding this research and to Dr Michael Redford and staff at the Marine Biological Association of United Kingdom for aiding with methodology checks and advice for research quality assurance.

References

[1] Bacon K L and Swindles G T 2016 Could a potential Anthropocene mass extinction define a new geological period? *The Anthropocene Review* **3** 208–17
[2] IPBES 2019 Summary for policymakers of the global assessment report on biodiversity and ecosystem services. S. Diaz et al. *(eds)* (IPBES secretariat, Bonn, Germany. 56 pages.
[3] Airoldi L and Beck M W 2007 Loss, status and trends for coastal marine habitats of Europe *Oceanography and Marine Biology: An Annual Review*
[4] Firth L B, Knights A M, Bridger D, Evans A J, Mieszkowska N, Moore P, O'Connor N E, Sheehan E, Thompson R C and Hawkins S J 2016 Ocean sprawl: challenges and opportunities for biodiversity management in a changing world *Oceanography and Marine Biology: An Annual Review* (CRC Press)
[5] Glarou M, Zrust M and Svendsen J C 2020 Using Artificial-Reef Knowledge to Enhance the Ecological Function of Offshore Wind Turbine Foundations: Implications for Fish Abundance and Diversity *Journal of Marine Science and Engineering* **8** 332
[6] Giakoumi S, Brown C J, Katsanevakis S, Saunders M I and Possingham H P 2015 Using threat maps for cost-effective prioritization of actions to conserve coastal habitats *Marine Policy* **61** 95–102
[7] Henry L-A, Mayorga-Adame C G, Fox A D, Polton J A, Ferris J S, McLellan F, McCabe C, Kutti T and Roberts J M 2018 Ocean sprawl facilitates dispersal and connectivity of protected species *Scientific reports* **8** 1–11
[8] Airoldi L, Beck M W, Firth L B, Bugnot A B, Steinberg P D and Dafforn K A 2021 Emerging Solutions to Return Nature to the Urban Ocean *Ann Rev Mar Sci* **13** 445–77
[9] Seaman W 2007 Artificial habitats and the restoration of degraded marine ecosystems and fisheries *Biodiversity in Enclosed Seas and Artificial Marine Habitats* Developments in Hydrobiology ed G Relini and J Ryland (Dordrecht: Springer Netherlands) pp 143–55
[10] Roberts C 2007 *The Unnatural History of the Sea* (Island Press)
[11] Knowlton N and Jackson J B C 2008 Shifting Baselines, Local Impacts, and Global Change on Coral Reefs *PLOS Biology* **6** e54
[12] Duarte C M, Agusti S, Barbier E, Britten G L, Castilla J C, Gattuso J-P, Fulweiler R W, Hughes T P, Knowlton N and Lovelock C E 2020 Rebuilding marine life *Nature* **580** 39–51
[13] Firth L B, Airoldi L, Bulleri F, Challinor S, Chee S-Y, Evans A J, Hanley M E, Knights A M, O'Shaughnessy K, Thompson R C and Hawkins S J 2020 Greening of grey infrastructure should not be used as a Trojan horse to facilitate coastal development *Journal of Applied Ecology* **57** 1762–8
[14] Rouse S, Porter J S and Wilding T A 2020 Artificial reef design affects benthic secondary productivity and provision of functional habitat *Ecology and Evolution* **10** 2122–30

[15] Hunter W R and Sayer M D J 2009 The comparative effects of habitat complexity on faunal assemblages of northern temperate artificial and natural reefs *ICES Journal of Marine Science* **66** 691–8
[16] Langhamer O and Wilhelmsson D 2009 Colonisation of fish and crabs of wave energy foundations and the effects of manufactured holes – A field experiment *Marine Environmental Research* **68** 151–7
[17] Drouin A, Archambault P, Clynick B, Richer K and McKindsey C W 2015 Influence of mussel aquaculture on the distribution of vagile benthic macrofauna in îles de la Madeleine, eastern Canada *Aquaculture Environment Interactions* **6** 175–83
[18] Le Gouvello R, Hochart L-E, Laffoley D, Simard F, Andrade C, Angel D, Callier M, De Monbrison D, Fezzardi D, Haroun R, Harris A, Hughes A, Massa F, Roque E, Soto D, Stead S and Marino G 2017 Aquaculture and marine protected areas: Potential opportunities and synergies *Aquatic Conservation: Marine and Freshwater Ecosystems* **27** 138–50
[19] Vasquez M, Allen H, Manca E, Castle L, Lillis H, Agnesi S, Al Hamdani Z, Annunziatellis A, Askew N and Bekkby T 2021 EUSeaMap 2021. A European broad-scale seabed habitat map
[20] Ellis J I and Schneider D C 1997 Evaluation of a gradient sampling design for environmental impact assessment *Environmental Monitoring and Assessment* **48** 157–72
[21] Methratta E T 2020 Monitoring fisheries resources at offshore wind farms: BACI vs. BAG designs *ICES Journal of Marine Science* **77** 890–900
[22] Ugland K I, Gray J S and Ellingsen K E 2003 The species–accumulation curve and estimation of species richness *Journal of Animal Ecology* **72** 888–97
[23] R Core Team 2019 *R: A Language and Environment for Statistical Computing* (Vienna, Austria: R Foundation for Statistical Computing)
[24] Wickham H 2016 *ggplot2: Elegant Graphics for Data Analysis* (Springer-Verlag New York)
[25] Wickham H, Averick M, Bryan J, Chang W, McGowan L D, François R, Grolemund G, Hayes A, Henry L, Hester J, Kuhn M, Pedersen T L, Miller E, Bache S M, Müller K, Ooms J, Robinson D, Seidel D P, Spinu V, Takahashi K, Vaughan D, Wilke C, Woo K and Yutani H 2019 Welcome to the tidyverse *Journal of Open Source Software* **4** 1686
[26] Anderson M J, Gorley R N and Clarke K R 2008 *PERMANOVA+ for PRIMER: guide to software and statistical methods.* (PRIMER-E Ltd.)
[27] Clarke K R and Gorley R N 2006 *PRIMER v6: User Manual/ Tutorial.* (PRIMER-E, Plymouth.)
[28] Smith J A, Cornwell W K, Lowry M B, Suthers I M, Smith J A, Cornwell W K, Lowry M B and Suthers I M 2017 Modelling the distribution of fish around an artificial reef *Mar. Freshwater Res.* **68** 1955–64
[29] Fabi G, Scarcella G, Spagnolo A, Bortone S A, Charbonnel E, Goutayer J J, Haddad N, Lök A and Trommelen M 2015 Practical guidelines for the use of artificial reefs in the Mediterranean and the Black Sea. *Studies and Reviews-General Fisheries Commission for the Mediterranean* **96** 1
[30] Lengkeek W, Didderen K, Teunis M, Driessen F, Coolen J W P, Bos O G, Vergouwen S A, Raaijmakers T, De Vries M B and Van Koningsveld M 2017 Eco-friendly design of scour protection: potential enhancement of ecological functioning in offshore wind farms: Towards an implementation guide and experimental set-up
[31] The Rich North Sea 2021 Linked in post, with footage: https://www.linkedin.com/posts/derijkenoordzee_monitoringsmissie-een-kijkje-bij-de-reefcubes-activity-6867073035120254976-ZX1T
[32] Ganias K, Perri R and Moditsis A 2021 Cuttlefish and squid egg deposition patterns on artificial devices and trap-like gears: implications for offspring survival and population management *ICES Journal of Marine Science* **78** 1614–22
[33] Grati F, Fabi G, Scarcella G, Guicciardi S, Penna P, Scanu M, Leoni S, Riminucci F, Frittelloni C, Gagliardini L and Bolognini L 2018 Artificial spawning substrates and participatory research to foster cuttlefish stock recovery: A pilot study in the Adriatic Sea *PLOS ONE* **13** e0205877

[34] Guerra Á, Hernández-Urcera J, Garci M E, Sestelo M, Regueira M, Gilcoto M and González Á F 2016 Spawning habitat selection by the common cuttlefish Sepia officinalis in the Cíes Islands (Northwest Spain) *Fisheries Research* **183** 44–54
[35] The Rich North Sea 2021 Offshore Test Site Project Infographic. Webpage. URL: https://de-rijke-noordzee-production.s3-eu-west-1.amazonaws.com/Website/CO5_Offshore_Test_Site.jpg, Accessed 13/01/2022

Appendix - Table A1. List of Reef Cube® clusters with inventory of Reef Cubes® per cluster type.

Table A1. List of Reef Cube® clusters with inventory of Reef Cubes® per cluster type.

Reef Site	Cluster type	Cluster Details
Reef 1	A	1.5m Reef Cube®
Reef 2	D	4 x 1m Reef Cube®
Reef 3	B	1.5m Reef Cube® containing 1 x 0.5m Reef Cube®
Reef 4	E	8 x 0.75m Reef Cube®
Reef 5	C	1.5m Reef Cube® containing 1 x 0.5m Reef Cube® and 3 x 0.15m Reef Cubes®
Reef 6	F	1.5m Reef Cube® with 70 x 0.3m Reef Cubes® scattered over.
Reef 7	D	4 x 1m Reef Cube®
Reef 8	E	8 x 0.75m Reef Cube®
Reef 9	B	1.5m Reef Cube® containing 1 x 0.5m Reef Cube®
Reef 10	A	1.5m Reef Cube®
Reef 11	C	1.5m Reef Cube® containing 1 x 0.5m Reef Cube® and 3 x 0.15m Reef Cubes®

AUTHOR INDEX